WINDBREAKS

WINDBREAKS

STEVEN BURKE

INKATA PRESS

1998

Acknowledgements

I thank the following colleagues and friends for their support, input and comments during the preparation of this book:

Lisa Richmond, Rob Youl, Rod Bird, Rowan Reid,
Jim Robinson, Tracy Jarvis, Ron Dodds,
Claire Dennis, Andrew Campbell, Roger Hall,
Phil Haines, David Connor, John Kellas

INKATA PRESS
A DIVISION OF BUTTERWORTH-HEINEMANN

AUSTRALIA   BUTTERWORTH-HEINEMANN 22 Salmon Street,
               Port Melbourne, Victoria 3207
SINGAPORE   REED ACADEMIC ASIA
UNITED KINGDOM BUTTERWORTH-HEINEMANN Ltd Oxford
USA   BUTTERWORTH-HEINEMANN Woburn, Massachusetts

National Library of Australia Cataloguing-in-Publication entry

Burke, Steven, 1960– .
Windbreaks.

Bibliography.
Includes index.
ISBN 0 7506 8951 X.

1. Windbreaks, shelterbelts, etc. I. Title.
(Series: Practical farming).

631.27

Edited by hexpress, Sydney
Typeset by Bookset, Type & Image
Printed by Chong Moh Offset Printing Pte Ltd

Contents

Table of Case Studies

Introduction

Shelter is a basic requirement of life. People, plants and animals thrive in sheltered environments. Creating a more sheltered environment is an important farm management objective. The only practical way to provide shelter for broadscale agriculture is by establishing windbreaks. Farmers and other land managers increasingly recognise the value of shelter, and windbreaks are consequently an increasing feature of rural areas in Australia and throughout the world. Windbreaks have a place on almost every farm.

Traditional rural landscapes in many parts of Europe consist of small fields with hedgerows or low windbreaks which delineate boundaries and control stock. Many of these landscapes date from the Middle Ages. More recently, development of agriculture in the heathlands of Denmark, the Great Plains of North America, and the steppes of the former Soviet Union about a century ago was associated with windbreak establishment. Wind is a dominant feature of the climate in these largely treeless areas.

In contrast, at the time of European settlement, most areas of Australia and New Zealand suitable for agriculture were largely covered by forest, woodland and shrubland. The compelling need of settlers was to create open areas for crops and pastures. Their challenge was to remove trees and shrubs from the landscape to allow for the expansion of agriculture. To some extent this resulted in the view that trees were antagonistic to agriculture. Clearing was undertaken with tremendous pioneering zeal. This created problems related to too little shelter and stabilising perennial vegetation. Ironically, the first windbreaks were established within a few years of land clearance. These were usually primarily to protect homesteads, and consisted mostly of introduced species. However, many farmers soon acknowledged the value of shelter to agriculture, and extended windbreaks along paddock margins, often making greater use of native species.

Smaller areas of naturally treeless plains occur in Australia and New Zealand, such as the Western Plains of Victoria and the Canterbury Plains in New Zealand. The agricultural development of these exposed regions was associated with the establishment of extensive windbreak systems in much the same way as had occurred in the treeless areas of the northern hemisphere.

Agricultural systems incorporating windbreaks are also becoming a feature of regions of many parts of the developing world, where the scarcity of land as a resource has resulted in the need to produce food and fibre requirements very efficiently. Integrating windbreaks with agriculture is often a very effective way of achieving this.

Almost every rural property has something to gain from the provision of shelter. Major benefits include increased crop and pasture growth, higher livestock productivity, improved horticultural returns, and reduced wind erosion, together with wildlife habitat and landscape improvement. To maximise sustainable farm productivity in many exposed areas, shelter is simply essential.

Once in place, a windbreak network begins to make a difference in just a few years and the benefits extend for generations. Windbreaks can be simply incorporated into most farms without major disruption to the primary enterprise, be it cropping, livestock or horticulture.

But just any old windbreak will not do. To maximise the value of windbreaks, careful attention needs to be given to their design, location, establishment and maintenance.

The first part of this book will describe the basics of how a windbreak works. This will be followed by a description of the productivity and conservation benefits they can provide. Finally, a guide will be provided to the best way to go about planning the establishment of your windbreaks. Some exciting new information will also be included from researchers about the increases in agricultural production that can be obtained by providing shelter, and from leading farmers about new and innovative approaches to windbreak design and layout.

CHAPTER 1

Windbreak Basics

How a Windbreak Works

To get the most out of windbreaks, it helps to understand how they work and what they can do for you. Understanding the fundamentals of how windbreaks work will provide the means for you to establish a windbreak network tailor-made to meet your precise requirements.

Windbreaks and Microclimate

Agricultural production in a region is controlled by many factors, of which climate is one of the most important. (This is the real reason why farming people talk about the weather all the time!)

We usually accept that nothing can be done about the climate, and most agricultural research and development efforts concentrate on improving cropping and animal production practices to increase farm productivity.

Windbreaks actually create small-scale changes in climate across the adjacent area. Localised climatic effects near the ground are sometimes called microclimate. These changes are important as it is the microclimate which agricultural plants and animals actually experience. Windbreaks have the capacity to alter microclimate dramatically. The localised effects of windbreaks on microclimate can be similar in significance to climatic differences between distant, agricultural zones.

Windbreaks influence most of the factors that comprise microclimate. A general view of these effects is shown on p.2.

First we will investigate those major characteristics of a windbreak which determine how it influences microclimate. This will be followed by a detailed description of the influence of windbreaks on each component of microclimate.

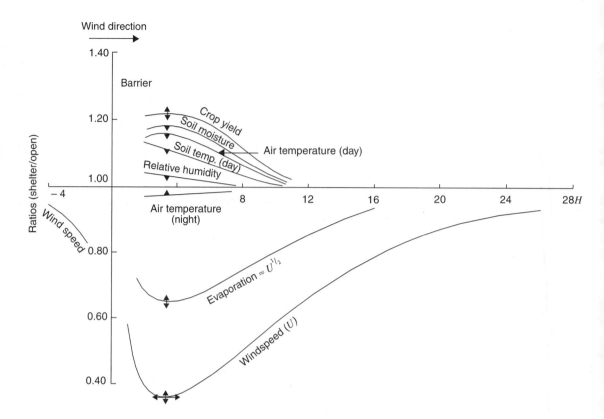

Figure 1.1 *Microclimate variation near a windbreak (after Marshall 1967); (H = height of the windbreak).*

Windbreak Characteristics

Windbreaks can be divided into two broad categories: natural and artificial. Natural windbreaks are strips or belts of trees, shrubs and grasses in various combinations and arrangements which are generally planted, although they can be remnants of the original vegetation. Artificial windbreaks usually consist of woven fabric, timber, metal or similar sheet material attached to a frame or fence. The vast majority of windbreaks established today are natural, largely because they can produce a broader range of benefits at lower cost. In some circumstances, however, artificial windbreaks are preferred, because they occupy less space, have no competitive effects, and can be erected almost instantly.

Most of the examples used in this book will be natural windbreaks. This is simply because they are the most common. Many of the benefits provided by natural windbreaks can also be provided by artificial windbreaks.

Two major factors affect the precise way in which a windbreak will affect the microclimate near it – these are height and porosity.

The height of a windbreak is obviously the distance from the ground to its top. For an artificial windbreak, height is usually uniform and simple to measure. For a natural windbreak, height can vary along their length and an average measure is usually used for any calculations.

Defining windbreak porosity is far more complex. The true porosity (sometimes called aerodynamic porosity) of a windbreak is the degree to which approach winds can pass through it. For thin, artificial windbreaks this is equivalent to the ratio of perforated area to total area. For natural windbreaks, determining porosity is more complicated. This is because a complex three-dimensional array of gaps determines their true porosity. No simple, direct technique exists to measure true porosity in the field.

The proportion of background visible from a position perpendicular to a windbreak is termed optical porosity. This can be determined approximately by visual assessment, or more accurately from black and white photographs. While optical porosity is a commonly used approximation of true porosity, they are really only equivalent for thin, artificial windbreaks. A natural windbreak with the same optical porosity as a thin artificial windbreak will always have a lower true porosity. Similarly, wider natural windbreaks have lower true porosity than narrower ones with the same optical porosity. Optical porosity is nevertheless frequently used to describe natural windbreaks due to its practicality and ease of assessment.

Windbreaks and their Effects on Wind

What is the wind? Wind is simply the movement of air over the earth's surface. This moving air behaves like water or other fluids, even though it can't be seen. Being mindful of the fluid nature of wind helps when trying to interpret its behaviour.

Wind is perhaps the least understood of the factors comprising microclimate.

The fundamental effect of a windbreak is to simply act as a barrier to wind. Windbreaks cannot stop the wind any more than a child playing with stones on a riverbank can dam the entire river. All a windbreak can do to wind is to deflect it.

Windspeed can only be reduced in one part of the flow by increasing it in another. Some of the wind approaching a windbreak is deflected at higher speeds over the top of it and some generally filters through. The result is a reduced wind speed near the ground on the leeward side.

Wind Speed

The most obvious effect of a windbreak is to alter the speed of winds near it. The pattern of wind speed changes is, however, far from simple.

The Influence of Windbreak Characteristics

The height and porosity of a windbreak largely determine how it alters wind speed. The extent of influence of a windbreak on wind speed is directly proportional to windbreak height (H). This means, for example, that if the height of a windbreak is doubled, the extent of its effect on wind speed (and other aspects of microclimate) is also doubled, and so on. For presentation and interpretation of wind speed data (and other microclimate effects), it is usual to measure the distance from the windbreak in terms of multiples of the windbreak height (H).

The distance at which wind speed is reduced to at least 80% of open field values is generally between 15 and 20 windbreak heights (15–20 H) (Caborn 1957), although extreme distances of up to 30 H have been reported. The effect of a windbreak on wind speed becomes marginal beyond horizontal distances of approximately 30 H leeward and 7 H windward (Bird 1991; Caborn 1957; Sturrock 1969 and 1972; Marshall 1967; Naegeli cited in Van Einerm et al 1964). Some slight wind speed reductions have been detected at distances in excess of 50 H.

Windbreak porosity strongly affects the pattern of wind speed variation within the shelter zone. Windbreaks of lower porosity produce lower minimum wind speeds (Caborn 1957; Sturrock 1969 and 1972). This means that the denser the windbreak, the greater the maximum degree of wind protection provided. In fact, the minimum wind speed behind a windbreak (expressed as a percentage of open field wind speed) is a good indicator of true windbreak porosity (Heisler and DeWalle 1988).

Early field research (for example that of Naegeli 1946) indicated that the rapid wind speed recovery behind less porous windbreaks could be such that they protected a smaller area. In other words a denser windbreak would have winds recovering to their open field levels over a shorter distance.

More recently wind tunnel studies (Raine and Stevenson 1977; Schwartz et al 1995) have enabled a better understanding of the effects of windbreak porosity on shelter. These studies show that this early work exaggerated the rate of wind speed recovery with high density windbreaks . The zone protected by denser windbreaks is in extent similar to, or in some circumstances even greater than, the protection zone offered by more porous windbreaks. In general, the zone with significantly reduced wind speed slowly increases as porosity decreases from 100% to 20%. Decreasing porosity below 20% does not further increase the extent of shelter. Several wind tunnel trials now indicate that the optimal extent of shelter is provided by windbreaks with a 20% porosity.

The extent of significant protection (say more than 20–30% reduction in wind speed) is similar for a wide range of windbreak porosities from medium to very dense windbreaks.

It is possible that the vertical distribution of porosity in windbreaks may influence the extent of wind speed reduction. Conjecture exists about which creates greater sheltered areas, uniformly porous windbreaks (Caborn 1965) or windbreaks which are more porous at ground level (Rosenberg et al 1983). In practice, it can be dangerous to allow excessive porosity at ground level, because increased wind speeds can be produced for a limited zone near the windbreak, potentially creating wind erosion and other problems.

The cross-sectional profile of a windbreak may also influence leeward wind speed recovery. There is some evidence (Caborn 1957) that windbreaks which are smooth or aerodynamically streamlined in cross section produce more rapid wind speed recovery than windbreaks which are rectangular in profile.

The general height of objects, such as crops and other vegetation, built structures and other obstacles above ground level is termed landscape roughness. Landscape roughness near a windbreak has an important influence on the effectiveness of the windbreak. Windbreaks shelter a greater area when approach winds are laminar, or smooth in nature, rather than turbulent. This is because a more turbulent approach flow increases vertical mixing of air in the lee, resulting in more rapid wind speed recovery. Greater landscape roughness near a windbreak reduces the sheltered area by increasing turbulence of the approach flow (McNaughton 1988).

Species	Belt type width	Height (m)	Trees/m of belt	Windward			Mid belt	Leeward					
				–3H	–1H	O		O	1H	3H	6H	12H	24H
1. Sugar gum	Sown, 20m, multiple row, not fenced	18	7	78	72	73	–	74	66	47	37	55	76
2. Sugar gum	Sown, 28m, four rows, multiple row, not fenced	19	4	85	76	76	–	107	76	49	39	72	91
3. Tuart	Single row, not fenced	15	0.43	95	81	–	108	–	110	73	61	77	100
4. Monterey cypress	Single row, fenced 4.5 m	10	0.29	89	80	–	56	–	58	37	30	54	82
5. Monterey cypress	Single row, fenced	15	0.17	86	61	–	92	–	80	50	27	54	(84)
6. Monterey cypress	Single row, not fenced	9	0.25	88	66	–	55	–	(49)	45	24	52	97
7. Monterey pine	Two rows, 6m, not fenced	20	0.33	62	47	45	–	45	40	37	45	74	95
8. Young acacias	Four rows, 7m, fenced	3	2	100	100	95	–	26	23	46	66	83	(91)
9. Red gum/ acacia	Six rows, fenced, 20m	6	2	97	87	–	–	76	61	35	43	70	91

(Table header span: Windspeed (% of open values) at various distances (H) from the belt)

This also explains why the distance protected by the second and subsequent windbreak in a parallel series is less than the first windbreak. The first windbreak reduces the approach wind speed for the second windbreak, but also increases the turbulence of the approach flow to it. This increases the wind speed recovery in the lee of the second windbreak. To provide the equivalent overall reduction in wind speed, the third windbreak needs to be closer to the second than the second is to the first.

Figure 1.2 *Windspeed reductions near a number of windbreaks in Victoria. (Data from Bird (1991) and Burke (1991).)*

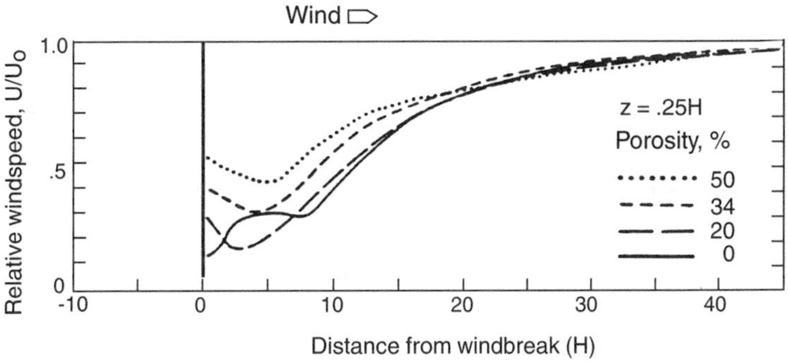

Figure 1.3 *Windspeed reductions (expressed as a proportion of open field wind speed) near windbreaks of different porosity. (From Raine and Stevenson 1977.)*

Greater atmospheric instability has a similar, though less significant, reducing effect on the area sheltered. This is also because under these conditions, approach winds tend to be more turbulent (Van Einerm et al 1964; Sturrock 1972).

The height of measurement above the surface greatly influences wind speed due to drag effects of the surface (Rosenberg et al 1983). It is therefore important to measure wind speed at a consistent height when assessing wind speed changes near windbreaks. Wind speed downwind of windbreaks as a proportion of open field wind speed is not greatly affected by height of measurement, if this is less than one third the windbreak height (Heisler and DeWalle 1988).

The Influence of Oblique Winds

Winds will rarely approach a windbreak from a perfectly perpendicular direction. In practice, wind speed reduction in the lee of windbreaks also depends on the approach angle of the wind. The two effects of oblique winds on wind speed near a windbreak are (per Heisler and DeWalle 1988):
• alteration of the area sheltered, and
• alteration of the effective windbreak porosity.

Oblique winds influence the area sheltered due to essentially geometric effects. As the approach angle of the wind becomes more oblique, points progressively closer to the windbreak experience full, open field wind speed as the wind misses the windbreak. The area sheltered reduces as this occurs (Burke et al 1988). Wind speed reductions are not greatly affected for winds less than 25° from the perpendicular. Also, the sheltered area is not reduced to zero with winds parallel to windbreaks. Even for winds parallel to natural windbreaks, momentum absorption of the tree crowns and drag effects on wind flow produce significant wind reductions. The zone with significantly reduced wind speed for winds parallel to natural windbreaks is approximately one-quarter (1/4) that for perpendicular winds. In other words, windbreaks have a substantial effect on wind speed near them regardless of the approach angle of the wind.

Windbreak performance can also be affected by oblique winds due to changes in effective porosity. The effect of oblique winds on aerodynamic porosity of two dimensional artificial barriers is, however, different to that of three dimensional natural windbreaks.

For artificial barriers consisting of vertical slats, wind speed in the sheltered zone increases as the approach angle of the wind becomes more oblique (Heisler and DeWalle 1988). This is because porosity increases with increasing obliqueness due to the increasing ratio of effective open area between the slats (wind can still be deflected through the gaps) to the area presented at right angles to the wind. Artificial barriers consisting of horizontal slats are less sensitive to changes in porosity with varying wind direction. Conversely, for natural windbreaks of significant width, porosity decreases as obliqueness of wind angle increases, since the effective width of the windbreak increases. The effectiveness of very porous windbreaks can be increased for oblique winds due to increased effective porosity.

Wind Turbulence

The movement of air across the earth's surface is rarely smooth. For some time it has been recognised that in addition to affecting mean horizontal wind speed, windbreaks also alter turbulence characteristics. Turbulent air flow exhibits eddying or rapid and irregular changes in wind speed and direction. Most importantly, the turbulent nature of wind is changed by windbreaks in terms of its eddy size and kinetic energy.

The zone immediately leeward of windbreaks has long been associated with reduced turbulence. More recently, sophisticated equipment has enabled the detection of distinct turbulence zones in the lee of windbreaks.

Wind deflected by a barrier is compressed and accelerated over it, creating a flow with higher turbulence. This turbu-

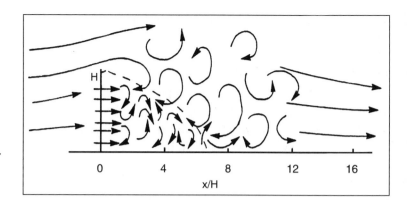

Figure 1.4 *Eddying patterns near windbreaks showing quiet and wake zones. (From McNaughton 1988.)*

lent flow then disperses downwind. A region of increased turbulence exists beyond a calm zone for a range of natural and artificial barriers under field conditions. The calm "quiet" zone of reduced eddy size and other turbulence characteristics occurs to about 8 H leeward of a windbreak for perpendicular winds, under stable conditions, and for low landscape roughness. This zone is triangular in profile, extending from the top of the windbreak to the quiet zone boundary at the surface. An extended zone of increased size and energy of turbulent fluctuations slowly reducing to open field levels exists downwind from this and is called the "wake" zone.

Turbulence effects are more pronounced for rigid artificial windbreaks than natural windbreaks. This is because wind acceleration and wind shear over the top of a rigid barrier is much greater than for a more flexible variable natural windbreak, creating greater downwind turbulence.

Taller barriers appear to produce a more extended quiet zone where there is high landscape roughness.

Windbreak porosity does significantly affect the extent of the quiet zone although it does influence other turbulence characteristics. Windbreaks with low porosity deflect more wind than more porous windbreaks, and produce greater turbulent velocity fluctuations at their top which results in increased turbulence in the wake zone, although lower minimum wind speeds are achieved. More porous windbreaks, by allowing more air to pass through, have higher mean wind speeds in the quiet zone; however, due to lower turbulent fluctuations at their top, wake zone turbulence is reduced.

There is no simple relationship between the wind speed and turbulence effects of windbreaks.

Multiple Windbreak Systems

Little research has been conducted on the effects of multiple windbreak systems on microclimate. This is despite the fact that ultimately, to optimise the benefits of windbreaks, multiple windbreak systems will have to be employed.

A few simple principles, however, are generally understood.

The first windbreak produces the greatest extent of wind speed reductions. Second and subsequent windbreaks have more turbulent approach winds, created by the first, reducing their effectiveness.

For uniformly spaced windbreaks, after the initial wind speed reductions of the first windbreak, the effect of the subsequent windbreaks is to maintain the level of wind speed reduction or to limit the rate of wind speed recovery. It is a common misunderstanding that second and subsequent windbreaks, at uniform spacings, can further reduce wind speed that has already been reduced by the first windbreak. A common observation for uniformly spaced windbreaks is a very gradual increase in average wind speed after subsequent windbreaks. This occurs until an equilibrium average wind speed reduction is reached, at a distance believed to be in excess of 50 H from the first windbreak (Heisler and DeWalle 1988).

Decreasing the spacing between windbreaks may only produce small additional wind speed reductions. It also appears that the porosity of the second and subsequent windbreaks is less important than that of the first.

Closely spaced windbreaks can produce continuous high quality shelter as they largely decouple the entire paddock from the general wind flow. The general flow skims over the top of the windbreak system.

Windbreak networks increase the overall landscape roughness, causing broadscale reductions in the wind speed. The atmospheric wind speed above the windbreaks may be reduced in addition to any known effects of wind speed reduction below the height of the windbreaks. This reduction is little understood or researched, but appears to extend the field of windbreak effect beyond the microclimate, and to a broader area or region.

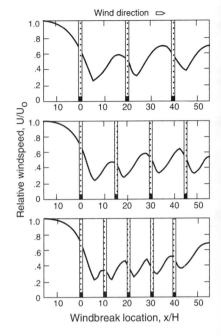

Figure 1.5 *Effect of spacing between windbreaks on wind speed for multiple windbreak systems. (From Naegeli 1964.)*

Other Components of Microclimate

By affecting the wind speed and turbulence, windbreaks also affect the movement of heat, water vapour and CO_2. This produces variation in temperature, relative humidity and CO_2 concentration.

Temperature, Relative Humidity, and CO_2 Concentration

Effects of Reduced Turbulent Mixing

The traditional view regarding the effect of windbreaks on temperature is that they increase the continental character of the climate, that is, they increase day temperatures and decrease night temperatures (Caborn 1957; Marshall 1967). Increases in daytime temperature of up to 2°C are widely reported to occur in a zone extending to about 10 H from a windbreak (Skidmore and Hagen 1970; Radke and Hagstrom 1973; Russell and Grace 1979; Benzarti 1990). Occasionally larger increases of up to 4°C have been reported (Brown and Rosenberg 1972, Burke 1993). The magnitude of daytime temperature increases is dependent on sensible heat advection, solar radiation (or sunshine), and surface wetness. Sensible heat advection is the transport of heat by wind, such as occurs on days of hot northerly winds. The less the sensible heat advection onto the site, the lower the solar radiation, and the wetter the soil surface, the smaller the increase. The greatest daytime warming near a windbreak therefore occurs on sunny days with hot winds when the surface is dry.

In addition to these daytime temperature increases, less pronounced reductions in temperature occur at night (usually up to 1°C) in a similar zone.

These temperature effects are due to reduced turbulent mixing in the calm leeward zone or quiet zone of a windbreak. Sensible heat generated at the soil or plant surface during the day, and temperature inversions created at night by heat loss by radiation from the surface (which then becomes a sink for sensible heat rather than a source) become trapped in the quiet zone. In effect, the quiet zone just behind the windbreak becomes "decoupled" from the broader environment.

In addition to these temperature effects, increases in relative humidity and slightly decreased daytime CO_2 concentration is found near windbreaks due to reduced turbulent mixing (the surface being a source of vapour due to evapotranspiration, and a daytime sink for CO_2 due to photosynthesis).

Wake Zone Effects

It has been speculated (McNaughton 1988) that the distinct turbulence zones behind windbreaks (the quiet and wake zones) may produce differing effects on temperature, humidity and CO_2 concentration. As turbulent transport of heat, vapour and carbon dioxide decrease in the quiet zone (creating the microclimate effects described above), they may increase in the wake zone. It is proposed that increased turbulent transport and vertical mixing in the wake zone may create opposite effects of a similar scale to those observed for temperature, relative humidity and CO_2 concentration in the quiet zone. This would mean that temperature and relative humidity would be decreased and CO_2 concentration increased in the wake zone. It should be stressed that these wake zone effects are mostly hypothetical, and are based on only a small amount of evidence (primarily temperature measurements). It may be that wake zone effects on microclimate are more significant for dense windbreaks which create a greater degree of wake zone turbulence than for more porous windbreaks.

Surface Temperature Effects

In addition to air temperature, windbreaks also strongly affect temperatures of plant surfaces such as leaves and stems. Surface temperatures of objects receiving solar radiation are directly related to wind speed. Wind reduces the boundary layer resistance (or the still air around surfaces) and therefore increases heat transfer from the surface. Where there is less wind, there is less heat transfer, and therefore the surface gets hotter. Shelter can consequently cause significant increases in plant surface temperatures. This effect is easily felt on our skin when we stand in the sun. Under still conditions, direct sunlight can become very hot; however, even a slight breeze on the skin can greatly reduce this effect. The effect of windbreaks on surface temperatures can then be very significant.

Soil Temperature

The temperature of the top few centimetres of bare soil is strongly influenced by the surface temperature and nearby air temperature. At greater depths, temperatures are buffered and soil temperature is more dependent on the relative extent of periods of daytime warming and cooling, and the magnitude of these effects (Guyot 1989). Over time, increased daytime air temperatures and soil surface temperatures (offset to some extent by any nighttime or wake zone reductions in air temperature) generally lead to increases in soil temperature. Reduced evaporative cooling in shelter also contributes to increases in soil temperature (Marshall 1967).

The magnitude of these increases is generally up to 3°C depending on depth of measurement, soil and atmospheric conditions (Marshall 1967; Van Einerm et al 1964). One study on the Canterbury Plains of New Zealand (Radcliffe 1985) detected increased maximum soil temperatures in sheltered zones of up to 4°C.

Solar Radiation

The direct effects of windbreaks on radiation balance due to shading can be significant immediately adjacent to a windbreak, but are of little consequence beyond 1 or 2 H (Marshall 1967). While a windbreak can cast much longer shadows than this, tall shadows, which are only cast when the sun is low and incident radiation is reduced, have a much less significant effect (Skidmore 1976). Windbreaks can also reflect radiation (Rosenberg et al 1983). North-south oriented windbreaks influence a smaller area than east-west windbreaks because shadows tend to be cast along the windbreak (Rosenberg et al 1983). Reflection of solar radiation by windbreaks facing the sun may contribute to increases in air temperatures near the windbreak.

Precipitation

Rain falling with wind is inclined, so that an obstacle such as a windbreak intercepting this casts a rainshadow (Guyot 1989). This can greatly affect the precipitation reaching the ground immediately adjacent to windbreaks. There is little reliable evidence, however, that a windbreak significantly influences rainfall away from a rainshadow zone of approximately 1 H. This zone can, however, extend further for wind-driven rain.

Many northern hemisphere cropping regions, in particular those using winter wheat, derive soil moisture benefits from redistribution of precipitation by snow drifting in the lee of windbreaks. "Snow farming" with windbreaks is important for successful cereal crop production in many severe-winter, continental climates with marginal precipitation (Black and Aase 1988).

Evapotranspiration and Soil Moisture

Soil moisture is influenced by a number of factors, of which evapotranspiration is the most important affected by shelter (ignoring redistribution of precipitation by snow drifting). Evapotranspiration is the combined effect of evaporation and water use by plants.

Water near the surface in bare soil will evaporate under most atmospheric conditions. Crop growth and development alters this by the additional effects of crop water use (or transpiration). To simplify the discussion, the influence of shelter on evaporation alone will be considered first, followed by a description of the effects on evaporation and transpiration combined, and finally, how these affect soil moisture.

Evaporation

The direct influence of wind on evaporation is evident in theory (Penman 1948) and from practice. Many field trials (Sturrock 1972; Skidmore 1976; Skidmore and Hagen 1970) confirm that evaporation is closely related to wind speed reduction. Evaporation is significantly reduced to a distance 10–12 H from a windbreak, and reductions have been measured to distances in excess of 30 H.

The increased humidity and reduced night temperatures reported in the quiet zone of a windbreak would reduce evaporation, while during the day, higher temperatures would tend to offset the effect of increased humidity. The relative magnitude of these effects on evaporation appears, however, to be small in comparison with the effect of altered wind characteristics in the lee of windbreaks.

Sensible heat advection (or flux) is the transport of sensible heat by wind. Reduction of sensible heat advection is an important shelter effect that reduces evaporation (Guyot 1989). This produces the strong drying effect which is particularly noticeable for hot dry winds. Up to one third of the energy used in evapotranspiration can be supplied by advection (Hagen and Skidmore 1974). Reduced sensible heat advection near windbreaks is particularly significant in dry continental climates (Guyot 1989), including Australia (Stern 1967). This is because dissipation of radiation as latent heat is less and therefore partition to sensible heat correspondingly greater.

Evaporation and Transpiration Combined

The transpiration component of total evapotranspiration is dependent on the degree of plant cover. When plant cover is small, evaporation comprises a major proportion of evapotranspiration and therefore similar moderating shelter effects to those described for evaporation would apply for evapotranspiration. When surfaces are not wet and/or plant cover is greater, transpiration becomes more significant and wind speed may increase, decrease or have no effect on total evapotranspiration depending on plant resistances to the diffusion of water vapour (Monteith 1964). Most field studies do, however, indicate an overall reduction in evapotranspiration in shelter (Rosenberg et al 1983).

Soil Moisture

Although it is difficult to distinguish the effects of evaporation from transpiration in developing crops, many field studies indicate that reduced evaporation in shelter has the effect of conserving moisture in the soil (Aase and Siddoway 1976; Lynch et al 1980; Sturrock 1981; Radcliffe 1985; Frota et al 1989; Ding and Zhang 1990).

The first Australian field study of soil moisture variation near windbreaks was conducted near Armidale in New South Wales (Lynch et al 1980). Soil water loss was significantly less (12.3 mm) from sheltered than unsheltered paddocks for a drying period of 29 days from field capacity. As the paddocks were grazed pasture, transpiration was minimised, so evaporation would form a more substantial proportion of total evapotranspiration. It was concluded that the evaporation component of total evapotranspiration was reduced, resulting in increased soil moisture. In a similar field study from a semi-arid area of north-eastern Brazil, Frota et al (1989) measured wind speed, pan evaporation and soil moisture to 0.5 m depth near a natural windbreak over periods of several months. Reduced wind speed corresponded with reduced pan evaporation and increased soil moisture (by up to 5.9%).

One of the few field studies of evaporation from bare soil (Aase and Siddoway 1976) is described in Figure 1.6. After irrigation, the soil surface remained wetter in shelter for about three days compared with exposed control areas. Consequently, the effect of the windbreaks on bare soils is to increase soil moisture for a period through reducing the rate of evaporation.

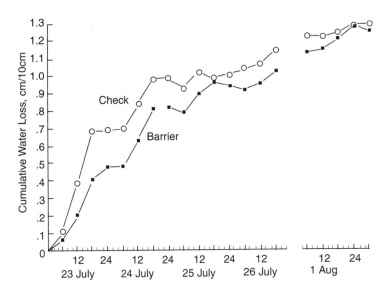

Figure 1.6 *Difference in mean cumulative water loss from bare soil between unsheltered and sheltered sites. (Aase & Siddoway 1976.)*

It has been pointed out (Rosenberg et al 1983) that the evaporation rate from wet, bare soil will be similar to the pan evaporation rate for only a few days, after which it decreases to zero over a period of time (the exact period depending on complex factors such as soil texture, hydraulic conductivity, and humidity). However, during the life of a crop, soil water is likely to be recharged by rainfall a number of times, repeating this moisture conservation effect. Frequent light rains which initially moisten only the soil surface have a better chance to infiltrate and be used by plants if evaporation is reduced in shelter. Better use of intermittent light rain in shelter may be very important at crucial periods of crop development, such as grain filling of cereal crops in spring. Where early crop and pasture growth and development in winter is slow, and rainfall occurs as small regular events (frequently the case in south-eastern Australia), soil evaporation may comprise 30–60% of total evapotranspiration (Siddique et al 1990). Enhanced soil moisture availability through reduced evaporation in shelter is consequently likely to be a significant shelter effect.

Water Use Efficiency

Evaporation, transpiration and soil moisture are interacting factors. Increased crop growth can confound the detection of reduced evaporation by soil moisture measurements. The leaf area of crops in shelter is often greater (Grace 1988a). Given that transpiration from an area of crop is a function of leaf area, sheltered plants which exhibit better growth may transpire more, resulting in no measurable differences in soil moisture. It is even possible that rapidly growing crops in shelter may dry soil more than in exposed sites (Carr 1972), even though soil moisture availability in the absence of a crop would be greater due to reduced evaporation.

By reducing evaporation as a proportion of total evapotranspiration, windbreaks increase productive water use (or the proportion transpired by plants). Water-use efficiency, defined as above-ground dry matter produced divided by evapotranspiration, can be useful in assessing the water conservation benefits of shelter for crops in these circumstances. Water-use efficiency is generally improved in shelter (Hagen and Skidmore 1974; Radke 1976; Rosenberg et al 1983).

Water Use by Natural Windbreaks

There are also negative effects of natural windbreaks on soil water availability due to extraction of soil water by trees and shrubs (Huxley et al 1989). Any increases in soil moisture away from a natural windbreak are offset to some extent by the competitive use of soil water by trees in the area immediately adjacent to it (Greb and Black 1961). This competitive zone usually extends to a distance of 1 H (Kort 1988), although it varies with soil type and tree species.

CHAPTER 2
Plant Productivity

Windbreak Benefits to Crop and Pasture Production

In this chapter we will first describe the fundamental responses of plants to the changes in microclimate created by windbreaks. This will be followed by a description of how these responses combine to affect overall crop and pasture productivity.

How Windbreaks Affect Plants

Plant growth and development is affected in a number of ways by the microclimate created by windbreaks. Here we will look at the most important plant responses.

Plant Water Relations

Soil Moisture Effects

We have seen that an important effect of windbreaks is to reduce evaporation and therefore increase soil moisture availability.

Water supply is the most critical environmental factor affecting crop and pasture yields in Australia. The high variability of rainfall in most of Australia's cereal growing regions largely explains the high year-to-year cereal crop yield variation. Drought at any time seriously affects vigorously growing plant tissue.

Improvement of soil water levels in shelter due to decreased evaporation has the capacity to reduce plant water stress, enabling plants to photosynthesise longer and greatly increase growth. For example, in the trial at Armidale in New South Wales (Lynch et al 1980), which found significantly greater soil moisture in shelter for a closely grazed pasture, there was higher herbage production as a result. In most of Australia, where soil moisture limits plant growth throughout most of the year, moisture conservation in shelter may be particularly beneficial. The Armidale study concluded that increased moisture availability for plant transpiration and growth can contribute greatly to plant productivity when moisture becomes limiting in late spring and summer.

Reduced evaporation in shelter is likely to have little effect on plants under conditions of abundant moisture availability and under severe moisture stress. Under intermediate conditions, plants in shelter can take advantage of relatively increased moisture due to reduced evaporation and can continue to transpire and grow, while exposed plants, suffering from moisture stress, have slow or no growth.

Direct Effects of Wind on Plant Water Use

Wind can also directly affect plant water use (or transpiration). John Grace from the University of Edinburgh has described two opposing effects of wind on transpiration. Firstly, wind decreases resistance to transfer of water vapour from the leaf, which tends to increase transpiration. Secondly, wind decreases leaf temperature, which reduces the vapour pressure gradient between the sub-stomatal cavities in the leaves and the atmosphere, decreasing transpiration.

Many have wrongly concluded that a major mechanism by which windbreaks increase plant production is by reduced plant water requirements thereby making better use of limited water availability. This is due to a common misconception that wind always increases water use by plants (or transpiration). Grace has described how the fundamental effect of wind is either to decrease water use or to not influence it. This basic effect can be complicated by water loss through physical damage of plant surfaces by wind. Lower transpiration with increased wind speed has frequently been observed (Rosenberg 1966; Drake et al 1970; Radke and Hagstrom 1973) although a wide variety of plant water use responses to shelter are reported from field trials (Rosenberg et al 1983).

Greater soil moisture availability due to reduced evaporation can clearly benefit plant growth. This is a most significant effect of windbreaks on plant production, particularly in dry regions. The variety of responses that have been reported in terms of plant water use in shelter indicates there are some differences in how plant species and varieties exploit this moister, more sheltered environment. Plant species which use water freely may derive greater benefit and exhibit more extreme responses to the increased moisture availability in shelter than those which are more water efficient. Regardless of this, almost all plants will benefit to some extent from the increased moisture availability induced by windbreaks.

Mechanical Effects

Wind can also affect plants directly by mechanical stimulation and damage of plant tissue.

Thigmomorphogenesis

The mechanical stimulation by wind exerts considerable influence on plant growth. Plants become shorter and thicker as a direct response to their movement by the wind. This reaction is termed thigmomorphogenesis. It probably occurs in all plants. The plant hormone ethylene is stimulated by mechanical stimulation which produces a decrease in stem elongation and increase in diameter growth. The increased growth of mechanical tissues at the expense of photosynthetic tissue reduces net plant growth. Thigmomorphogenesis appears to produce plants more capable of withstanding strong winds and therefore may be an evolutionary adaptation. Thigmomorphogenesis also appears to protect plants from frost injury.

Thigmomorphogenesis is a major reason why seedling plants grown in a glasshouse are soft and fleshy and often need to be "hardened off" outside to develop more strength prior to planting out.

It is probable that reduced thigmomorphogenesis is part of the explanation for the more luxuriant plant growth commonly reported in sheltered environments. By limiting plants' exposure to wind, windbreaks may decrease or eliminate thigmomorphogenesis, thus improving plant productivity. However, there is a possibility that even small amounts of mechanical perturbation may produce the maximum thigmomorphogenic response (Grace 1988a); if this is so, the significance of windbreaks in counteracting thigmomorphogenesis may be small.

Physical Damage

Windbreaks can reduce direct physical damage to plant surfaces caused by wind. This includes microscopic rupturing of the leaf epidermis that impairs the capacity to control water loss, and more apparent damage such as abrasion and tearing, resulting in loss of leaf area.

Other more obvious forms of plant damage by wind can also be reduced by windbreaks. Seed can be lost or excessively buried by soil movement, affecting seed germination and plant emergence. Sandblast can be severe after plant emergence. This can be a major problem in some parts of south-eastern Australia, reducing crop yields (Cooke et al 1988). It has been reported in New Zealand that lodging (flattening of crops due to high winds) has been reduced by windbreaks (Sturrock 1981).

The quality of crops can also be improved by providing shelter which will reduce physical damage. This is particularly important for sensitive vegetable and speciality crops (Baldwin 1988).

Temperature Effects

The effect of the small increases in daytime air temperature near windbreaks on the rate of photosynthesis or carbon fixation of plants is likely to be small. This is because the rate of photosynthesis is not a highly temperature dependent process within the range of temperature variation generally experienced.

It is likely that the effects of temperature variation on cell division and extension in the meristematic zones are more significant than changes in rates of carbon fixation. Increased cell division and extension results in an increase in leaf area. An increase in the rate of production of leaf area will increase dry matter production, given a constant rate of net photosynthesis (Russell and Grace 1979).

Important differences have already been described between plant surface temperatures and air temperatures in the lee of windbreaks. Plant surface temperatures can be greatly reduced by the direct effects of wind. Sheltered plant surfaces shaded by plant leaves and other plant tissue may be only marginally warmer than those in exposed plants (reflecting air temperature differences), while sheltered plant surfaces in full sunlight may be much warmer. Increased rates of cell division and leaf growth in particular will be created by increased plant meristem surface temperatures near windbreaks.

Temperature also has a strong influence on plant development as opposed to growth. Plant development is the progression from one life stage to the next. It is important to differentiate between plant development and growth when assessing the effects of temperature. There is generally a direct linear relationship between plant development and temperature. In other words, as temperature increases, changes from one life stage of a plant to the next will occur more quickly. Plants generally germinate from seed, emerge, develop and mature faster at higher temperatures.

The effects on plant growth and crop yield of advancing plant development due to the temperature effects of shelter may be positive or negative. A possible negative outcome is that plant development at higher temperatures may outpace increases in physiological processes such as photosynthesis.

In some circumstances, it is possible that windbreak temperature effects may be sufficient to cause temperatures to exceed the tolerance limits of a plant. Extreme temperatures above or below these tolerance limits can cause irreversible physiological damage. Plant injury could be caused either by excessively high temperatures in hot climates due to daytime warming, or frost damage due to nighttime cooling near windbreaks.

The temperature effects of shelter are likely to be an important cause of plant response in many situations. Temperature modification is one of the most important windbreak benefits in cold and temperate parts of the world, although it may be detrimental in hot climates.

Other Effects

Several other microclimate factors influenced by windbreaks may also affect plant growth.

Soil Temperature

Shelter-related variation in soil temperature probably affects levels of root respiration and root growth. The soil temperature effect of shelter on plants is likely to be small compared to the soil moisture effect, although it may play a role in seed germination and early growth.

Carbon Dioxide

Carbon dioxide concentration directly affects plant photosynthesis. However, the small effects of shelter on atmospheric CO_2 concentration are unlikely to significantly affect crop response (Grace 1988a; McNaughton 1988; Rosenberg et al 1983).

Radiation

Obstacles, such as trees in windbreaks, can affect the quantity and spectral distribution of solar radiation. Shading may contribute to depressed growth of some plant species to a distance of 1–2 H (Marshall 1967). In hot climates, beneficial effects of shading have also been reported. The most important factors influencing patterns of radiation reaching the surface (Reifsnyder 1989) are:
- row orientation and spacing
- windbreak height and
- crown structure.

Relative Humidity

Moderate variation in relative humidity has little effect on plant growth at normal atmospheric levels (Salisbury and Ross 1978). It is therefore unlikely that the variation in relative humidity induced by windbreaks directly influences plant growth.

Field Crop and Pasture Production

We have seen that windbreaks can affect plant growth and development in numerous different ways. These individual plant responses combine and interact to determine the overall effect on crop and pasture yields.

This net effect, simply stated, is that shelter benefits plant growth. John Kort, a researcher from Indian Head, Canada, has reviewed worldwide more than 50 studies undertaken over a period of more than 50 years. Kort found the overwhelming majority of research indicates overall positive responses to shelter for a wide variety of crops.

Numerous trials have demonstrated that beyond a narrow zone of reduced growth immediately adjacent to windbreaks, large increases in yields (often in the range 20–30%) can extend to a distance of 10–12 times the height of the trees.

Year-to-Year Variation in Crop Response

Crops can exhibit considerable variation regarding the effect of shelter on plant growth. Yield responses to shelter frequently vary from year to year. The precise prediction of a particular crop response to shelter is fraught with uncertainty. The major reasons for this variation are:

- varying crop responsiveness to shelter
- the varying nature of shelter
- rare strong winds which cause damage obscuring the effects of prevailing winds, and
- varying limiting factors to agricultural production for different regions.

Crop yield variation in shelter is a combined response of a number of variable processes which interact with each other and the developing crop.

Responses for Different Climatic Zones and Crops

The pattern of plant growth near windbreaks is likely to vary from region to region and from crop to crop.

Crop Species

Kort also found that some crop species were more responsive to shelter than others. Overall, he found that spring wheat, oats and corn responded moderately; rye, barley, winter wheat, hay and millet better; and lucerne responded greatly. High responsiveness of forage crops may be explained by a common observation that the vegetative response to shelter is greater than reproductive yields. Crops highly sensitive to those microclimatic factors which windbreaks strongly affect (most notably moisture availability) are likely to be the most responsive, particularly if that factor is strongly limiting.

Crop	No of field years	Mean yield increase (%)
Oats	48	6
Spring wheat	190	8
Corn	209	12
Hay	14	20
Winter wheat	131	23
Barley	30	25
Millet	18	44
Lucerne	3	99

Relative reponsiveness of crops to shelter (fom Kort 1988). Based on 50 studies around the world conducted between 1932 and 1985. Field years are the number of fields measured multiplied by the number of years measured.

The nature of the plant response is also different for different climatic zones. While increases have been observed from a wide variety of climates, the processes of plant response may vary.

Next we will consider some of the crop and pasture responses that have been observed in different broad climatic zones.

Temperate, Seasonally Dry Climates

It has been frequently observed that crop responses to shelter are higher in drier seasons and in drier regions.

The most comprehensive Australian field study on the effect of shelter on agricultural productivity was conducted by CSIRO researchers at Armidale in New South Wales. Sheep were grazed in enclosures protected by windbreaks over a five-year period. As shelter was provided by metre-high artificial barriers, this study did not consider the competitive effects of natural windbreaks on production. Significantly more pasture production occurred in shelter, with the largest differences occurring in dry periods. Measured in terms of the metabolisable energy intake of grazing sheep, pasture production was increased by 18% over five years at the highest stocking rate. The increased pasture growth due to shelter was believed to be a more important cause of the increased livestock production than the direct effect of reduced wind on the animals. The positive effect on pasture growth was caused by significantly higher soil moisture levels due to reduced evapotranspiration in shelter.

The CSIRO study at Armidale concluded that significant increases in productivity can potentially be derived from the establishment of shelter in the seasonally dry agricultural areas of Australia.

More information for shelter effects on pasture is available from the seasonally dry Canterbury Plains of New Zealand. Radcliffe (1985) made the first attempt in New Zealand to quantify the effect of windbreaks on pasture production. Dry matter was measured at points 0.3, 3, 5 and 12 H from a windbreak. Pasture production was 60% greater at both 3 H and 5 H than at either 0.3 H or 12 H. It was inferred that soil moisture was the most significant direct cause of increased production. This increased level of pasture growth in shelter would allow for a greater stocking level to be maintained, or a reduced requirement to provide supplementary feed.

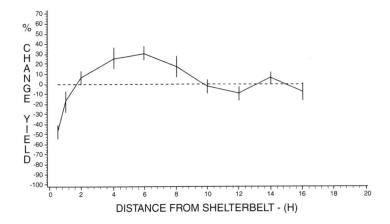

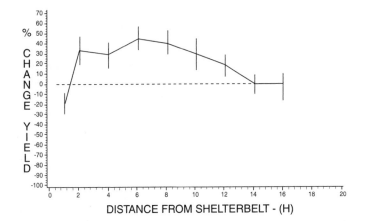

Figures 2.1 and 2.2 *Crop yields near windbreaks in 1990 at the Rutherglen Research Institute, being respectively wheat above, and oats below. The distance is in terms of the number of windbreak heights from the windbreak. (From Burke 1991.)*

There is now strong evidence from two sites in southern Australia which also indicates, at least in these areas, that shelter can improve crop yields. Red gum and acacia windbreaks at Rutherglen in Victoria increased wheat yields by up to 30%, and oat yields by up to 40%. These increases occurred in a zone starting about 1.5 tree heights from the windbreak, and extended to at least 10 times the height of the trees.

The Rutherglen study found that higher yields correlated with increased soil moisture and increased daytime temperatures. Increased moisture availability, particularly during grain filling in spring, was considered to be the dominant effect.

This study demonstrated that crop yield benefits on a site could be achieved on both sides of a windbreak, and for windbreaks of different orientations. The maximum yield increases were achieved on the south side of an east–west

Lupin and Oat Yields Increased by Windbreaks

Gary and Jan English farm on the sandplain north of Esperance, Western Australia, which receives about 450 mm rainfall per annum. Four parallel north–south oriented windbreaks were planted about 200 m apart in 1984 as a trial to control erosion of these light soils. The windbreaks consisted of three rows of radiata pine, with the rows within the windbreak 3 m apart. Of particular interest at this stage are some of the early crop yield responses. Some yield increases (up to 10%) were measured for oats, although lupins produced the most spectacular yield results. In summary these were:

- an increased lupin yield of between 19% and 22% when the area of the windbreaks is included, and
- a net yield increase of 27% for the lupin crop area between the windbreaks.

Overall, the windbreaks were a profitable investment in terms of the crop yields alone.

These crop yield benefits are, however, only part of the story. The windbreaks are also expected to provide a range of other benefits. Long-term returns from timber are expected from the trees in the windbreak. These are expected from posts, poles, and sawlogs. "In addition, the wind erosion hazard has been effectively controlled by reduction in wind speed only five years after planting", Gary said.

Figure 2.3 *Gary and Jan English – Esperance, Western Australia.*

oriented windbreak. The protection from drying northerly winds in spring was thought to have been particularly important.

Further information comes from a study conducted by Western Australian Department of Agriculture researcher David Bicknell. The two-year trial was carried out on the property of Gary English in the Esperance region of Western Australia, which has a strongly Mediterranean climate. Lupin yield increased by up to 40% and overall by an average of 20.5% (including the areas taken up by the windbreaks) between *Pinus radiata* windbreaks. Oats on the same site showed less response.

In another trial in New Zealand (Sturrock 1981), oat yields were increased by an average of 35% between one and six tree heights from a windbreak, with the maximum increase of 51% occurring at a distance of four tree heights. Surface soil moisture decreased with distance from the windbreak and better moisture availability was considered a significant factor.

Little information is available from similar climatic zones elsewhere in the world. One study from a Mediterranean climate was conducted in Italy, and recorded an 18% increase in the yield of spring wheat from 2–10 H (Van Einerm et al 1964).

These trials generally indicate the importance of the effect of windbreaks on soil moisture availability, and how wind depresses plant growth due to its drying effect. This windbreak effect is likely to be more significant in drier climates.

We have seen in Chapter 1 how evaporation is directly related to wind speed, and how evaporation behind windbreaks is greatly reduced. Reduced evaporation means that the rain that falls can be more effectively used. When the soil is wet, less is lost through evaporation, allowing a greater proportion to infiltrate and become available to plants for growth. Light showers which would normally evaporate quickly will be of greater benefit to crops and pastures. In temperate, seasonally dry climates in Australia, New Zealand and elsewhere, plant growth is generally limited by moisture availability, and windbreaks can result in increased crop and pasture growth as a result.

Continental Climates with Severe Winters

A high proportion of the agricultural land in the northern hemisphere has a continental climate. Precipitation occurs through rain and snow, and some moisture redistribution occurs by snow drifting. This is particularly so in Northern China, the steppes of the former Soviet Union, and the Great Plains of the United States and Canada. John Kort

Figure 2.4 *A well-established native windbreak protecting a cereal crop in a dry year near Roseworthy, South Australia.*

(Kort 1988) has compared, for a number of studies, yield increases in shelter to 30 H in areas with snowy winters (20.8% for 377 field years) with yield increases where winters are snowless (12.5% for 313 field years). It appears that snow accumulation by windbreaks is a significant effect in this climatic zone.

A large body of work indicating positive wheat responses to shelter is available from the continental climate of the Great Plains of North America. A typical response comes from a study undertaken by the University of Nebraska (Brandle et al 1984) for winter wheat in eastern Nebraska, where yields increased by 14.8% from 0.5 H to 5 H over 18 field years. This study included an economic analysis, which found a positive net present value and a 15-year pay-back period for the investment.

The warm zone which develops during the day near windbreaks can assist in the cool season growth and development of crops and pastures where temperature is limiting, particularly in these severe cold climate zones. As soil moisture is still frequently highly limiting throughout much of this zone, increased moisture availability is also likely to be a significant windbreak effect here.

Kort also summarised spring wheat yield responses over 116 field years near windbreaks from a number of research sites on the northern Great Plains of the United States and Canada. This shows an average yield increase of 3.5% in the zone to 15 H on each side of the windbreaks, including the

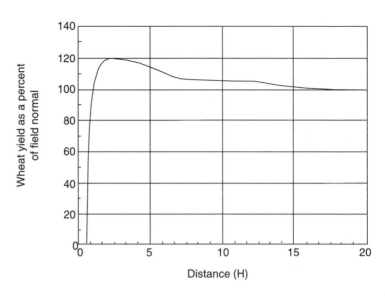

Figure 2.5 *Average 116 field year wheat yield near windbreaks in North America. (Kort 1988.)*

zones of reduced crop growth and the area taken up by the windbreaks themselves.

A relatively small number of studies report marginal or negative responses to windbreaks in the Great Plains of North America. One study in North Dakota (McMartin 1974) found spring wheat declined by 8% to 13 H over 14 field years. Although a small positive influence on paddock yields was detected, the competitive effects of the tree rows produced the overall negative result. While this is a response rarely reported in the literature, it does demonstrate the negative effects of tree competition. This varies mainly with tree species (in this case, the highly competitive Siberian elm, *Ulmus pumila*) and soil type.

Temperate Maritime Climates

A number of studies have come from middle latitude, temperate maritime climates, especially in Germany and Denmark. Despite being less susceptible to excessively dry conditions, increased crop yields in shelter are also widely reported. One example (in Van Einerm et al 1964) recorded a mean yield increase of 9.2% for oats to 30 H.

Tropical Climates

Some information is available from semi-arid tropical areas in India. In Andhra Pradesh, Reddi et al (1981) found windbreaks greatly increased the yields to a distance of 20 H of ground nuts (by 39.6 to 43.1%), pigeon peas (by 38.5 to 46.9%) and pearl millet (by 23.3 to 63.6%). Yield increases were attributed to lower evaporation and transpiration resulting in greater soil water availability. In Gujerat, Vora et al (1982) found a 24% yield increase for spring wheat, and a 13% increase for mustard for unspecified distances. Increases in plant height, whole plant weight, weight per grain, and total number of grains were detected.

A study undertaken on the Atherton Tablelands in tropical northern Australia (Sun and Dickson 1994) found that windbreaks increased overall potato yield by 6.7% (including the land taken up by the belts). Significant increases in potato size and proportion of A grade potatoes were also detected. As this site was irrigated, it was concluded that reduced wind damage and temperature effects were the most important windbreak influences on yield.

It is possible that in many tropical areas, where soil moisture and temperature are not as limiting to plant growth, windbreaks will provide fewer microclimatic benefits to plant growth than in cooler and drier climates. Windbreaks are still frequently planted in these areas for other benefits such as timber production, fodder production, and soil improvement.

Some Negative Effects on Crop and Pasture Growth

Trees can also reduce crop or pasture growth in a narrow zone immediately adjacent to them. This zone usually extends from one half to one windbreak height. It may also be greater or less depending on the extent of the competitiveness of the trees, and the sensitivity of the crops to this competition. It is rare to see a competitive zone extend beyond 2 H. In some circumstances, the competitive zone may be minimal or non-existent. This may occur, for example, under irrigation, where the water applied negates the competition for moisture from the tree, or in tropical areas where the benefits of shade to the crop exceed the competitive effects (for example with coffee crops).

The primary reasons for this competitive zone beside windbreaks are shading and interception of rainfall (described in more detail in Chapter 1), combined with root competition between the crops and the trees for moisture and nutrients. In some rare instances trees can also produce chemicals which inhibit plant growth near them. This effect is called allelopathy. There are some examples of this being caused by windbreaks (Singh and Kohli 1992).

The extent of competition between trees and crops or pastures is largely dependent on canopy structure and root distribution of the trees. The growth characteristics of the tree species are consequently very important. The zone of reduced crop or pasture growth can extend further with particularly extensive shallow-rooted species, while with deep-rooted species, this is much reduced. Excessive competition between the trees and crops or pasture can be minimised by ensuring that the trees are established into deep ripped lines and by regular deep ripping between the trees and crops along the paddock boundary.

This negative effect is, however, generally small when compared with the beneficial effect of shelter further from the trees.

Some authors (McNaughton 1988; Radke 1976) have also suggested that increased production is confined to the quiet zone of wind turbulence in the lee of windbreaks, and that there may even be a negative response in the wake zone. No substantial evidence exists in the literature to support this. The overwhelming consensus of the field trials undertaken around the world indicates no apparent negative influence in the wake zone, and that enhanced crop yield extends well beyond the quiet zone.

Horticultural Production

Horticultural crops (vegetables, speciality crops, orchards and garden plants) respond to shelter in the same fundamental way as other plants. In addition to this, horticultural crops derive the following additional benefits from the provision of shelter:

- Reduced rubbing damage. Damage to fruit skin is caused by rubbing against adjacent fruit, plant stems, leaves and trellises. This causes downgrading of crop quality which has been estimated to be as high as 30% for mangoes and citrus in southern Queensland. There is a strong link between the quality of shelter and the quality of the crop for most wind-sensitive fruit and vegetable crops.
- Decreased wind damage to orchard trees, vine canes and other plants. Mechanical damage to leaves and stems from wild movements by the wind can be significant. Form damage due to wind is also common for tree and vine crops. For example, Macadamia nut trees in northern New South Wales, southern Queensland and Hawaii are susceptible to this damage due to brittle wood, and benefit greatly from the provision of shelter.
- Minimised abrasion damage due to soil movement. This is particularly important for early-season vegetables grown on mobile sandy soils.
- Improved pollination efficiency.
- Better targeting, control and timing of spraying operations, and reduction of harmful spray drift.
- Earlier maturation of crops. The microclimate changes created by windbreaks (particularly increased daytime air and surface temperatures) accelerate crop development. Earlier maturing of horticultural crops often produces a price advantage for growers.

These combined effects of improved crop quality and quantity provide a strong economic incentive for the establishment of elaborate shelter systems for many horticultural

Figure 2.6 *An excellent windbreak with medium-dense porosity sheltering a high value macadamia nut orchard near Lismore, New South Wales.*

crops. Price of horticultural crops is strongly linked to quality. One study in New Zealand (McAneny et al 1984) showed exportable kiwifruit yields declining from 40 kg per vine near a windbreak to less than 10 kg per vine in more exposed locations, with fruit damage increasing from less than 5% to greater than 40%. These enormous benefits are reflected in investments in excess of $15 000 per ha to establish both natural and artificial windbreaks in prime kiwifruit orchards.

The benefits to horticultural crops are usually reported to extend to 10 H with the maximum benefits occurring between 3 H and 6 H. The yield benefits commonly range from 5–50%.

Timber and Other Tree Crop Production

As well as enhancing production from adjacent crops and pastures, windbreaks can also directly provide timber and other tree crops without excessively compromising shelter benefits. At the end of the life of a windbreak, the task of renovating or replacement can be expensive. It makes economic sense to plan to have a valuable harvestable crop at this stage. The alternative is generally an expensive task of renovating or replacing a senescing old windbreak, at best salvaging some low quality timber.

Firewood, posts and poles, and sawlogs may be produced for on-farm use or sale to supplement and diversify farm income. Because the precise harvest time for timber is not so critical, timber returns can be taken when returns from other farm enterprises are low, thereby acting as a financial buffer.

Growth rates of timber trees in windbreaks are often higher than for the same species in plantations. This is because of more fertile sites and lower levels of competition. In addition, harvesting costs of windbreak trees can be lower than for plantations, due to better access. With good establishment and management, the quality of the timber produced can be very high, making marketing easier and commanding a higher price.

The incorporation of timber trees into windbreaks can enable the economically viable establishment of timber tree species due to the combination of timber, shelter and other benefits, where the production of timber alone from woodlots is uneconomic. In the majority of dryland farming areas in Australia with less than 500 mm rainfall, it is unlikely that many native timber species will be planted economically as plantations in the near future. Timberbelts can, however, be a profitable investment in these areas.

To produce usable timber, care must be taken with species selection, windbreak design and management. These issues are discussed in the greater detail in Chapter 8 "Windbreak Design" and Chapter 10 "Maintenance".

CHAPTER 3

Animal Production

Extremes in climatic conditions can have enormous effects on the productivity of farm animals. Numerous studies from Australia and overseas indicate that the provision of shelter improves livestock performance through moderation of some of these effects.

We have already seen how shelter can improve pasture production, which in turn provides benefits to livestock. More available feed means that animals will be heavier and in better condition, or that higher stocking rates can be maintained.

But as well as providing more feed, windbreaks can offer other important benefits to animal production.

Prevention of Death From Exposure

The majority of livestock in Australia graze in the open year round, and they may consequently become susceptible to harsh climatic conditions from time to time.

Cold, wet and windy conditions frequently combine to produce conditions lethal to animals in the field. Wind chill is the effect of increased cooling experienced because of the combined effect of wind and low temperature. This occurs because wind increases the loss of body heat from the surface of an animal. This situation becomes even worse when the animal is wet, and heat losses due to evaporative cooling also occur. Animals in the open under these conditions can be extremely vulnerable.

Perhaps the most widely recognised benefit of shelter to livestock in southern Australia is the prevention of the death of newborn lambs from exposure in wet, cold, windy weather. Most lamb losses occur within three days of birth. Lambing

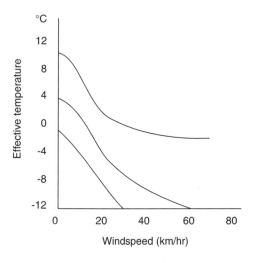

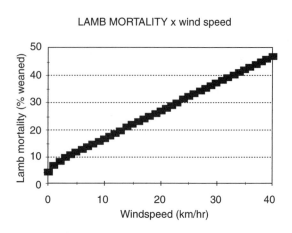

percentages can be very low in many regions in Australia because of this problem. It has been estimated that in southern Australia, on average, as many as 15% of newborn lambs die from exposure. Several Australian studies have shown that lamb mortality in cold, wet weather can be at least halved by the provision of adequate shelter. Lamb losses can be commonly reduced from 20% to 10% by the provision of adequate shelter and stock management.

Of the three factors which create exposure problems in livestock (cold, wet and wind), wind is the most practical to moderate through the provision of shelter.

The greatest reductions in mortality from exposure are produced by windbreaks designed to provide significant wind reductions when conditions of extreme exposure arise. Areas of high quality shelter created near dense windbreaks are required to achieve the maximum benefits. Alternatively, stock can be confined in paddocks with natural shelter (due to topography or dispersed shelter across the paddock such as tussock grasses or shrubs).

A chill index model has been developed (Donnelly 1984) which relates temperature, rainfall and wind speed to lamb mortality. This shows that by reducing wind speed from 10 m/s to 2.5 m/s through the provision of shelter, lamb mortality is at least halved.

Ewes do not always seek shelter when lambing, and natural sheep camping and lambing areas are often on exposed high ridges. Unshorn ewes are less likely to seek shelter to lamb than shorn ewes. To minimise lambing losses, either dispersed shelter must be available across the paddock, the

Figures 3.1 and 3.2 *Effects of wind speed on effective temperature (from Reid and Stewart 1994), and reduced lamb mortality with lower wind speed (Donnelly 1984).*

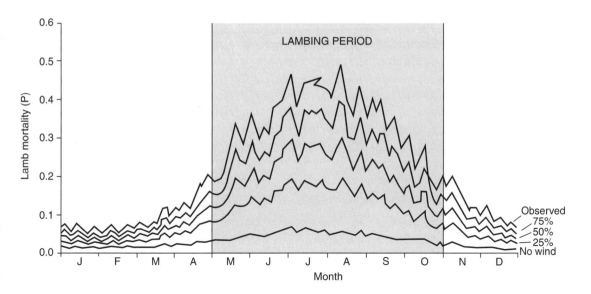

Figure 3.3 *The effect of reducing wind speed on the probability of lamb mortality throughout the year. (Donnelly 1984.)*

natural lambing sites must be sheltered, or the ewes must be confined to a "lambing haven", a confined area of high quality shelter. Good management is important to ensure lambing ewes are in the right place at the right time. Set stocked lambing paddocks need to contain enough feed to last the lambing period, often 6–10 weeks.

Similar benefits have been demonstrated for newly shorn sheep. These are generally susceptible to death from exposure for two weeks following shearing. The condition of the animals at the time also greatly affects their ability to survive these stresses. A graphic example of this occurred on 2 December 1987 in south-western Victoria where over 100 000 newly shorn sheep were lost. Well-sheltered farms in the middle of the worst affected areas escaped with few losses, where farmers were on hand to move sheep into sheltered areas. A similar event occurred in the south-west of Western Australia in January 1982 when an estimated 100 000 sheep also died. Lower losses than these examples occur virtually every year in most regions.

If high quality shelter is available for susceptible stock, it is important to plan ahead to ensure the ability to move them quickly to these areas in bad weather. It can be worth changing the location of fences to encourage stock to move to shelter in bad conditions. The location of shelter is crucial. It is often best to locate areas of maximum shelter in the downwind corners of paddocks or on high ground because sheep are often driven by high winds and it can be very difficult to move sheep into a gale.

Other areas where animals are confined, such as feedlots and stockyards, should also be given high priority for protection by high quality windbreaks. Sheltering these areas will reduce feed requirements, improve the weight and condition of stock, and result in improved animal health.

Figure 3.4 *Susceptible off-shears sheep sheltering in a gale. A livestock haven is a small confined area of high quality shelter. (Photo Jim Robinson.)*

Although cattle are much more hardy than sheep, calves can still be susceptible to exposure, especially when newborn. On average, in Kansas in the USA, a 2% increase in calf survival is gained from the provision of windbreaks (USDA 1994). Pigs and goats are extremely susceptible to death from exposure, even more so than sheep. This is because they do not have skin covering with the same thermal properties as greasy sheep wool. Shelter should be a high priority for these animals.

In warmer regions, provision of shade should also be considered to reduce stock losses (particularly for calves and lambs) from heat stress.

Improved Efficiency of Feed Utilisation

Less dramatic, though equally valuable, is the greater productivity of stock grazing in sheltered conditions. Productivity of farm animals is optimal for a comfort zone of a relatively narrow temperature range. Animals experiencing

Laneway Shelter a Safe Haven

Neil and Sue Lawrence have established over the years a network of laneways across their 1300 ha Balmoral property in western Victoria. The laneways are a tremendous labour saver when moving stock around the property. In some areas, around the house and shearing shed, red gums have been established by direct seeding to provide shelter to the laneways and surrounding areas.

A cold southerly snap hit in December 1987 during shearing when 1600 off-shears sheep were held in an exposed paddock near the Lawrence's house. During the wild night, they moved the sheep into a sheltered laneway. They lost 23 sheep in the 200 m before reaching shelter, but all those that made it survived. "In the laneways it was as though a switch had been turned off a big fan. The sheep stopped dying and some even started to graze."

While the Lawrence's flock escaped almost unscathed, one neighbour lost 600 off-shears sheep and the district total was 30 000. Neil reckons that the $200 or so that it cost him to direct seed those trees was the best investment he ever made.

Figure 3.5 *Neil and Sue Lawrence – Balmoral, western Victoria.*

excessively hot or cold conditions require more energy to maintain basic metabolism and thus have less energy available to increase body weight or to produce meat, milk or wool. In other words, more feed is required simply to counter the environmental stress. Under adverse conditions, livestock performance is inefficient relative to the quantity of feed eaten.

Experiments with penned sheep and cattle have shown that strong wind and rain double the energy requirement of animals for maintenance. One Australian study from the New England Tablelands in New South Wales showed that cold stress can depress sheep liveweight gain by 6 kg, and can depress wool growth by 25% (Lynch and Donnelly 1980). A study in Montana in the USA showed that beef cattle protected by windbreaks were on average 16 kg heavier than those in unsheltered feedlots (USDA 1994) It is well known that dairy production is improved by providing sheltered conditions. In southern Victoria it has been calculated that the provision of shelter can increase milk production by 30%. Ten per cent of this is due to greater efficiency of conversion of feed, and 20% due to the greater amount of feed available (Fitzpatrick 1994).

Grazing behaviour, and consequently feed intake, can also be affected. This may be partially compensated for by increased grazing when the conditions moderate. Animals under adverse environmental conditions generally graze less, and sometimes not at all.

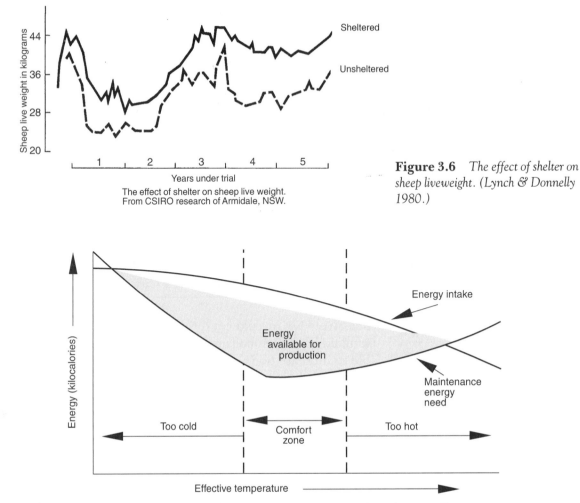

The effect of shelter on sheep live weight.
From CSIRO research of Armidale, NSW.

Figure 3.6 *The effect of shelter on sheep liveweight. (Lynch & Donnelly 1980.)*

Figure 3.7 *Energy required for animal maintenance and energy intake at different effective temperatures demonstrating the effects on energy available for animal production. (Anderson 1987.)*

Figure 3.8 *Sheep grazing on a well-sheltered pasture in Victoria's western District.*

Intensive Shelter for Dairy Production

Simon Park sees a number of benefits flowing from the establishment of an intensive shelter system from his 120 ha Wonthaggi dairy property. Simon has developed a long-term whole farm plan which features paddock subdivision and the planting of a regular series of windbreaks on the paddock boundaries. The windbreaks are generally parallel and spaced from 50–100 m apart, producing a very intensive shelter system.

Windbreaks are being established by a combination of direct seeding and planting of seedlings.

Simon is convinced of the increased milk yields that can result from sheltered grazing conditions for his herd. The farm is also near the coast and very windswept, so increased pasture growth can be expected from the provision of intensive shelter. While this is a long-term investment, Simon is convinced it will pay off.

"I also expect to see benefits from the natural control of pasture insects from increased bird life", says Simon. "More birds and other wildlife on the farm to me says that there is a healthier ecological balance."

Figure 3.9 *Simon Park – Wonthaggi, Victoria.*

Shelter trees can also provide substantial additional benefits due to the provision of shade. Where these benefits are likely to be substantial, shade plantings within paddocks should be considered in addition to windbreaks.

Shade can also make a valuable contribution to animal production, particularly in hot areas, such as the semi-arid and tropical north of Australia, but seasonal benefits can also be obtained in temperate regions.

Excessive heat depresses the condition of cattle by reducing feed intake. Heat stressed cows produced fewer calves and the calves produced have a lower birth weight. Calves frequently die of heat stress, particularly in tropical northern Australia. Heat stress can also markedly affect dairy production. One Queensland study (Davison et al 1988) showed that the provision of shade increased milk production for each cow by 2 kg per day, combined with an improvement in milk composition. European cattle breeds benefit more from provision of shade than tropical breeds.

For sheep, excessive heat is detrimental to ram fertility, and in ewes can reduce ovulation rate and conception. Heat stress in sheep can also lead to death, particularly for lambs. Ewes will assist their lambs to seek shade if this is available (Bird et al 1984).

In summary, the provision of adequate shelter can prevent dramatic stock losses under extremely adverse conditions. It can also provide small regular returns due to improved animal productivity. Shelter therefore produces a combination of benefits to livestock. In the case of sheep, this combination includes increased feed availability, reduced lambing losses, reduced off-shears losses, and increased efficiency of conversion of feed to meat and wool. In southern Victoria it has been estimated that these factors can combine to produce an overall increase in sheep productivity of 29% (Fitzpatrick 1994).

Figure 3.10 *Rob Batters – St Arnaud, north central Victoria.*

A Haven For Sheep and Pasture

Rob Batters farms about 1000 ha near St Arnaud in north central Victoria where he breeds stud Dorset sheep, lot feeds beef cattle, and crops about 20% of the area. Rob believes the open, treeless nature of the area means that the establishment of trees and productivity go hand in hand.

In 1988 Rob set aside a 9 ha area for a "warm paddock" of maximum shelter using an alley farming layout. Bays of phalaris based pasture 30 m wide were left between a series of three-row windbreaks comprising bull-oak (*Allocasuarina luehmannii*), river she-oak (*Casuarina cunninghamiana*) and black wattle (*Acacia mearnsii*). This was established at minimal cost as the windbreaks were unfenced. "These tree species were selected because they are nitrogen fixers and do not compete strongly with pastures", he said. Consequently, pasture growth has been excellent right to the base of the trees.

For the first four years during the establishment of the trees, the area was cut for silage. Rob feels that the silage yields were greatly improved by the provision of shelter. In the dry year of 1990 when no other areas where suitable, the paddock yielded 15 t/ha of silage. In 1991 and 1992 about 25 t/ha were cut. Sheep were first introduced to the area in the winter of 1993 when feed was short. Rob has also noticed a distinct improvement in the soil structure in the treed area.

Enthused by his success, Rob now plans to extend alley farming to other areas on the property.

CHAPTER 4

Shelter and Farm Economics

Windbreaks and Business

We have seen that windbreaks can increase crop and live-stock productivity on farms. Increased productivity does not, however, necessarily indicate that windbreak establishment is a wise investment. Most farmers will want to know whether a decision to plant windbreaks will be financially sound. Whether this is so depends on a wide range of factors, including:

- the effect of the windbreaks on agricultural production
- the area of the farm devoted to windbreaks
- the design, layout and growth characteristics of these windbreaks
- whether there are returns from timber or other tree crops, and
- the costs of windbreak establishment.

In general, agricultural returns which accrue over many years must be traded off against the short-term costs of windbreak establishment and the area used for windbreaks.

To understand the effect of this multitude of factors on the economic returns from investing in windbreak establishment, several computer models have been produced. These allow the user to test the effect of varying these factors on the financial benefits produced. Design and management options can be tested to identify the windbreak system which will produce optimal economic results. The various models tend to specialise in particular regions or systems. Several provide useful insights into windbreak economics. They are a very important tool to assist landholders in tailoring the most profitable windbreak system.

Dr Rod Bird of the Pastoral and Veterinary Institute near Hamilton in Victoria has calculated the economic effect of

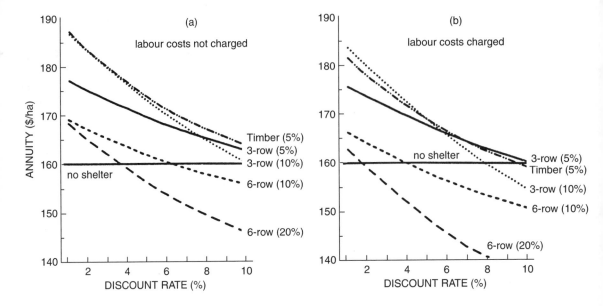

Figures 4.1 and 4.2 *Annual economic returns from an investment in three- and six-row windbreaks 250 m and 500 m apart on a model 400 ha farm in western Victoria. (Bird 1991.)*

providing shelter for a model 400 ha sheep farm in this area. The costs and benefits from planting three- and six-row windbreaks either 500 m or 250 m apart (these options covering from 5% to 20% of the farm) were compared with providing no shelter at all. The annuity or net returns per year are compared in Figs 4.1 and 4.2 for a range of discount rates. These results indicate that devoting 5% of the farm to windbreaks will increase returns under every scenario tested. When 10% of the farm is devoted to windbreaks, profitability depends on the discount rate, and whether or not the farmer provides the labour. The model indicated that it is more profitable to have 10% of the farm under narrow windbreaks 250 m apart than wide windbreaks at 500 m apart. Having 20% of the farm under windbreaks only proved economically viable for low discount rates. Timber production from the windbreaks can further enhance profitability, particularly if farm labour costs are not charged.

Peter Bulman of the South Australian Woods and Forests Department has produced a simple model for the dryland agricultural areas of South Australia.

This model indicates that in a 450 mm rainfall area producing crops and livestock, the establishment of a standard 10 m wide three-row native windbreak protecting an adjacent 10 ha area produces an economic benefit–cost ratio of 3.3. In the same situation the establishment of a 30 m wide sugar gum (*Eucalyptus cladocalyx*) timberbelt produces a benefit–cost ratio of 4.5.

Another more sophisticated model for personal computers, called *Farmtree*, has been developed in Victoria by Bill Loane of the Department of Natural Resources and Environment. This assesses the profitability of a wide range of investments in trees on farms including windbreaks. The model was developed predominantly for Victoria, although it is certainly relevant to other parts of Australia. The user can specify and vary a large range of site factors and management strategies, and *Farmtree* calculates the economic returns achieved. By varying key factors and re-running the program several times, an indication of the optimal design and management approach can be determined.

While a wide range of factors can affect the economics of windbreak establishment, all computer models show that windbreaks can be economically very attractive for a wide range of designs and management approaches. Generally, well-designed windbreaks are a good investment.

These models are very useful in choosing the options which will maximise profitability. The effect of altering a number of windbreak design and layout factors on overall financial returns can easily be investigated.

The extent of the competitive zone can have a significant impact on overall financial returns. This is because the competitive zone extends both sides of the windbreak and there is a wide variation in competitiveness of different tree species. The area with depressed growth near a windbreak can range from minimal to quite extensive. Consequently, the use of non-competitive species and management practices which minimise tree/crop competition are highly recommended.

Farmers are frequently concerned about giving away too much land to windbreaks and minimise their width. Interestingly, moderately increasing windbreak width generally only has a marginal effect on profitability. Hence the addition of one or two rows to increase windbreak performance or to produce more timber is often justifiable. Windbreak width more significantly affects profitability on high productivity agricultural lands where the returns from agriculture are large. In these areas, the addition of a substantial number of rows can, however, begin to take a toll on overall profitability (as seen by doubling the number of rows in the data from western Victoria above). In areas where profitability of agricultural and timber production is similar, windbreak width has less effect on overall profitability.

Windbreaks and Profitable Dairy Production

Alan and Judy Billing have used the *Farmtree* computer model to demonstrate that wind-break establishment is profitable for them on their Colac dairy farm.

In 1994 Alan and Judy established a 110 m long three-row windbreak consisting of shining gum (*Eucalyptus nitens*), blue gum (*Eucalyptus globulus*), and radiata pine (*Pinus radiata*).

The *Farmtree* model suggests that pasture production in the area behind the belt would increase by 10% once the trees reach 30 m, combined with a 7% improvement in animal performance. Because of the costs of establishing and managing the trees and the fact that land is taken out of production, the Billings lose money until year nine when pruning is completed and shelter benefits are increasing.

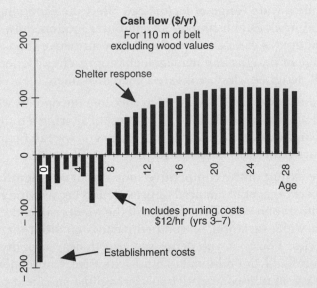

Figure 4.4 *Cash flow anticipated from the Billings' windbreak investment*

Ignoring interest rates, the project breaks even in year 14. If the Billings had to borrow money at 5% real value (ie 5% plus inflation) then the break-even point is not until year 18. Shelter alone is providing about $110 a year from the 110 m of belt after the 18th year. At age 30, the trees are estimated to be worth $3000/110 m or nearly $110 per tree.

Nothing is guaranteed in farming, especially when growing a timber crop which takes more than 20 years to mature.

Alan and Judy recognise the need for shelter and have planted trees for this purpose alone for many years. The option of a commercial timber return is seen as a bonus. The only additional cost is in the thinning and pruning of the windbreak. On the idea of harvesting a windbreak, Alan's response is simple: "I'll be planting more trees on the farm, so by the time I come to harvest any, the whole farm will be sheltered".

by Rowan Reid

Figure 4.3 *Alan and Judy Billing – Colac, Victoria.*

The distance between windbreaks also greatly effects the profitability. Narrow alleys between windbreaks (say, less than 100 m) produce earlier shelter benefits, but crops and pastures here suffer later due to the competitive effects of the windbreaks, and because there is insufficient space to allow the full benefits of increased agricultural production to develop. For very large distances between windbreaks, the economic returns from windbreak establishment becomes small. Often windbreak spacings of between 150 m and 300 m provide optimal economic benefits.

The discount rate (interest rate less inflation rate) used has a significant effect on economic returns. Lower discount rates greatly increase the profitability of windbreak investment. This occurs because the present value of long-term benefits of windbreaks such as timber and long-term shelter are greatly reduced by high discount rates. The effect of discount rates on returns can be clearly seen from Dr Rod Bird's study in western Victoria described above.

A rapid growth rate to maximum windbreak height tends to increase profitability, while actual windbreak lifespan is less important. This is because the effect of most commonly used discount rates is to significantly reduce the net present value of long-term increased agricultural productivity from a long lived windbreak, while early shelter benefits created by fast growth have a major positive effect on net present value.

The overall economic benefits from the provision of shelter are also determined by the nature and degree of crop or livestock response, the base unsheltered economic returns from agriculture, and the value of production from timber or other tree crops.

In general, these approaches to understanding the economics of windbreaks ignore all the other important long-term and non-financial benefits of trees on farms. These benefits include land protection, provision of wildlife habitat and landscape benefits. When these are taken into account, the argument for tree establishment becomes even more compelling.

CHAPTER 5

Conservation

The Environmental Benefits of Windbreaks

Wind Erosion Control

Australian soils are generally extremely old, fragile and susceptible to erosion. Soil is the fundamental resource on which agriculture is based and its condition largely determines the sutainability of our farming systems. It is non-renewable within normal agricultural planning horizons. Wind erosion is a major threat to the sustainability of agriculture in the semi-arid lands of Australia, where soil loss generally is far greater than soil formation. Agronomic improvements have to some extent masked the effects of soil loss on productivity of croplands in Australia.

The wind speed required for soil to begin to move is determined by soil type, pasture or trash cover, soil surface condition, soil structure and moisture. Soils are particularly prone to erosion when they are loose and dry (sandy soils being particularly erodible). Soil surfaces which are smooth and have little or no vegetative cover, surface trash or organic matter are more susceptible. Traditional cultivation practices associated with ley farming significantly expose soil to wind erosion. The size of the paddock or erodible area is another important factor. Large unbroken areas are more erosion prone than smaller areas.

Fine fractions are lost before the heavier particles and these contain the bulk of the soil nutrients and organic matter. Very small soil particles are easily transported into the air and moved in suspension for long distances. A small amount of wind erosion can lead to a significant reduction in soil fertility. More extreme examples of this result in dust storms. These are a common sight in summer and autumn, particularly in the cropping regions of southern Australia. Much of the expense and effort of fertilising land is lost and wasted through wind erosion. A single wind erosion event can literally mean thousands of dollars of soil nutrients being lost from the farm. In one study in the red-earth mulga country of central Queensland, it has been estimated that the erosion of the top 10 cm of topsoil would result in the loss of 450 kg of phosphorus and 1080 kg of nitrogen per ha (Boylen 1983).

Heavier soil fractions move along in little hops or creep along the soil surface. These are often deposited downwind along fencelines, bordering vegetation, or in ditches and channels.

Soil movement increases greatly as wind speed increases above a certain threshold. For soil in an erodible condition, such as loose, sandy soils, wind erosion rate above this threshold is proportional to the cube of wind speed. This means that a doubling of wind speed will result in an eight-fold increase in erosion. Even small reductions in wind speed near windbreaks can greatly reduce the movement of soil by wind.

The effects of wind erosion can be reduced in three fundamental ways.

The first is through crop and soil management practices which produce a less erodible soil surface condition. These conservation farming practices such as minimum tillage, trash retention, cover crops, and controlled grazing are widely recommended by extension services and are commonly used in many areas. Significant reductions in wind erosion have been achieved in Australia and elsewhere through the widespread adoption of these techniques. Wind erosion is, however, still a major problem, particularly in drought years, due to persistent use of traditional cultivation or imperfect application of conservation farming techniques. Recommended minimum tillage practices can still result in soil loss in extreme wind erosion events.

The second approach is to reduce the wind itself through the establishment of windbreaks. The wind speed reductions achieved by windbreaks within 10 H (windbreak heights) are generally more than sufficient to reduce the speed of a perpendicular wind to below the soil movement threshold. The precise effect of a windbreak on soil movement will depend on the unsheltered wind speed, the thresholds for the movement of soil particles (largely dependent on soil type and surface condition), and the design of the windbreak.

The third approach is to break up and isolate the areas of erosion prone soil. The aim of this is to reduce the continuous length of erosion prone soil in the direction of eroding winds. This can be achieved with tree and shrub windbreaks or, alternatively, by perennial or annual grass or crop barriers, strip cropping, or alternate row cropping. Perimeter windbreaks isolate paddocks from each other, and within-paddock windbreaks (such as alley farming systems) can be used to break up continuous erodible areas within paddocks.

Windbreaks can consequently reduce wind erosion in two ways. Firstly by reducing the wind speed, and secondly by breaking up and creating more discontinuous areas of erosion prone soil.

Careful attention to windbreak design and location is important to maximise the benefits from wind erosion control (see Chapter 8 on windbreak design).

The establishment of windbreaks can make a valuable contribution to controlling wind erosion. They contribute to a more reliable, permanent solution to wind erosion by reducing complete reliance on correct annual farm management decisions, particularly under drought or other difficult conditions.

A combination of practices incorporating the strategic use of windbreaks is frequently the best strategy to minimise wind erosion.

Windbreaks can also assist in soil improvement as well as conservation. Trees and shrubs improve soil nearby by the incorporation of organic matter from leaf and twig fall, and by trapping wind-blown nutrients. Deep-rooted vegetation such as trees and shrubs can transport small amounts of nutrients from weathering subsoil to the surface. More important for agricultural land may the cycling of leached applied fertiliser and nitrate from nitrogen-fixing plants from greater depths to the surface. Some trees and shrubs such as acacias, casuarinas and allocasuarinas are themselves nitrogen-fixing and may return small but useful amounts of nitrogen to the soil. In many developing countries, the use of branch prunings as green manure is an important method of soil improvement.

Water Erosion and Salinity Control

Land degradation problems requiring the treatment of whole catchments, such as sheet and gully erosion and some forms of dryland salinity, can be aided by strategic establishment of windbreaks as a part of an overall catchment treatment program. The introduction of a moderate level of tree and shrub cover, through the establishment of windbreaks or other agroforestry systems can provide many of the protective benefits of the original vegetation cover, while maintaining or enhancing agricultural production.

It is ironic that in Australia, where moisture availability usually limits agricultural production through much of the year, so many land degradation problems are caused by an inability of crops and pastures to use enough of the incoming rainfall. Infiltration of unused rainfall through the soil profile to the groundwater is the major cause of dryland salinity. Excess runoff causes water erosion. On average an estimated 5–10% of rainfall is unused in Australian agriculture and percolates into the groundwater or runs off (Peck and Hurle 1973).

The water used by trees is significantly greater than by agricultural crops and pastures. It has been estimated that 30–40 mature trees per ha are sufficient to use all of this excess rainfall. Establishing trees and other deep-rooted perennial plants on farms helps produce more closed water and nutrient cycles, retaining rainfall and nutrients that are applied to that site and using them productively.

Establishing windbreaks is one approach to achieving increased water use in a way which maintains or increases agricultural production. Important additional methods include other agroforestry systems, retaining remnant vegetation, and use of improved agronomic practices and crops which use more water and increase productivity.

Windbreaks on sloping land can effectively intercept, and slow down excess runoff. The trees and shrubs in windbreaks extract moisture from the soil profile (either infiltrating rainfall or water direct from the water table). Traditionally, trees for salinity control have been targeted to areas of high groundwater recharge. These are generally areas high in the landscape (hilltops and ridges) with shallow, stony soils. Hydrological research in Victoria has recently shown that belts of trees at the "break of slope" between hill country and flatter land can be particularly beneficial for increasing water use.

The protective canopy formed by trees reduces the direct impact of raindrops on the soil surface which break up and disperse soil particles. Tree leaf litter (and green manure

Fighting Erosion and Salinity with Alley Farming

In 1975 Noel Slee stood on the high point of his property at Marnoo in Victoria's Wimmera and watched a rainstorm wash off tonnes of topsoil in a river of red into the Richardson River. Some of his best land was also being lost to salinity along the river. Resolving that there was no long-term future in farming in this way, Noel has subsequently switched to a number of conservation farming practices including minimum tillage cropping, pasture improvement, and more recently the establishment of areas of alley farming.

In a salinity prone area, Noel's aim is to drop the water table below the current critical level. To achieve this, several three-row windbreaks have been established, consisting of timber species red gum and spotted gum, together with lower shelter species and saltbush for fodder. Noel plans to use low cost electric fencing to protect the windbreaks from sheep. Improved pasture in the alleys consists of lucerne, phalaris and "Balansa" clover.

"The improved pasture already has had the effect of lowering the water table", Noel said. "We hope that the windbreaks will further help in the process of saving land that was very nearly lost from any form of agricultural production."

Figure 5.1 *Noel Slee – Marnoo, Wimmera, Victoria.*

crops) can protect the soil surface and increase the organic matter content, increasing the soil stability and resistance to erosion.

In many developing countries, population pressure has necessitated the planting of crops on steeper, more water erosion-prone slopes. The introduction of hedgerow or agroforestry based systems in these situations can greatly minimise proneness to soil erosion. One example of this is the use of leucaena hedgerows on the contour in combination with the establishment of crops such as corn in hill country in the Philippines. The hedgerows lead to the formation of natural terraces and also provide green manure, fodder, and fuelwood.

Tree establishment can also assist in the control of streambank erosion beside rivers and creeks and reduce the extent of landslips on steep land.

Where tree planting is recommended to help control a specific land degradation problem (for example, an eroding gully), the planting can often be designed to also provide shelter benefits.

Landscape and Amenity Value

Trees and shrubs are important in influencing the appearance and character of our rural areas. Enjoyment of the landscape can be greatly enhanced by trees. They can introduce naturalness into artificially created agricultural environments. Of course, improving the appearance of a farm generally also means improving its capital value.

Natural landscape values are generally enhanced if windbreaks follow natural lines in the landscape such as ridges, contours and streams. The unsympathetic regimented use of straight windbreaks, without reference to natural landforms results in an obviously created landscape. It is relatively easy to introduce curves and bends in windbreaks so that they are more in harmony with the natural landform. The use of local or indigenous species also helps to retain the natural character of an area.

Windbreaks can be used to hide undesirable views or "wall off" spaces for privacy (for example, the homestead). This can greatly increase the sense of security and isolation. In addition, visually offensive features or undesirable views can be screened. Alternatively, attention can be directed to a particular point or scene through lines or avenues of trees and shrubs.

Unwanted noise can be absorbed and reduced by windbreaks. Vegetation reduces noise levels by diffusing and absorbing sound, particularly high frequency sounds. Dense, tall windbreaks are best for noise reduction. The closer to the sound source the windbreak is located, the more sound reduction is experienced behind the windbreak. The effectiveness of windbreaks as sound barriers is increased by the incorporation of earth mounds or other solid barriers.

In densely settled urban areas, belts of trees provide a buffer between incompatible land uses, and minimise adverse affects of noise, air and visual pollution

Fire Protection

It is commonly believed that trees in rural areas generally add to fire risk. Observations of fire behaviour in recent years clearly indicate, however, that strategically placed, well-established and maintained windbreaks can be an asset in controlling and minimising the damage caused by fire.

Windbreaks provide fire protection by:
- Reducing the wind speed and thus the speed and intensity of fire. A small increase in wind speed results in a much greater increase in the rate of spread of a fire. The message from this is simple — the use of windbreaks to reduce wind speed will enable the speed of the fire front

to be reduced to the point that fire fighting crews are able to work directly on the fire front. An effective windbreak offers the only practical means of reducing wind speed and therefore reduces the rate of spread of a fire.

- Catching burning airborne material, or deflecting it up over the windbreak so it is carried well to the lee or down-wind. This protects buildings and other assets from flying embers.
- Providing protection from radiant heat.
- Deflecting heat and smoke-laden winds.

CSIRO research indicates that the location of trees near fire-breaks greatly reduces their effectiveness. This is because if the trees catch fire they can easily shed sparks and burning embers across the firebreak. For this reason, important strategic firebreaks should be located well away from wind-breaks and other trees. A firebreak immediately in front of a windbreak may reduce the chance of the windbreak itself catching fire, which may be desirable, particularly if it is a fire sensitive species. This firebreak should not, however, be relied on as a major fire control line.

Trees should also be kept well away from power lines to min-imise fire risk.

Buildings and open spaces in the lee of windbreaks (but not directly against them) can be protected to various degrees from the effects of fire. These areas can also provide a life saving refuge for stock. Cases have been observed where windbreaks have assisted saving these assets in a fire. Maximum protection is provided about 3–5 H away from the windbreak.

Careful attention to windbreak design and orientation is important to maximise fire protection benefits. Think about the likely directions of fire approach and the major assets you wish to protect.

It has been recommended to break up long sections of wind-breaks to ensure long grass or accumulated fuel beneath does not act as a "fuse". This phenomenon has, however, rarely been observed, and gaps may actually aggravate the problem in some circumstances by funnelling and increas-ing wind speed. Reduction of fire fuels along the edge of windbreaks is generally recommended.

Wildlife Habitat

Windbreaks can increase opportunities for wildlife simply by providing greater habitat diversity in the rural landscape.

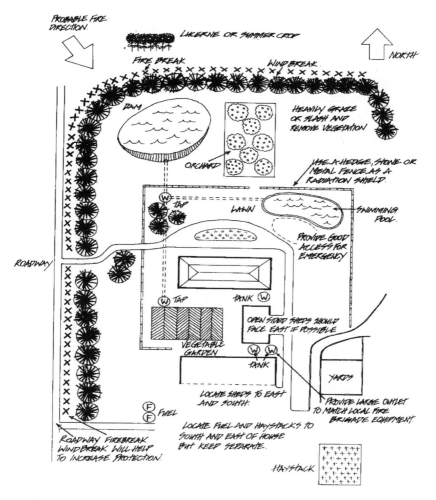

PROBABLE FIRE DIRECTION.

LUCERNE OR SUMMER CROP

NORTH

FIRE BREAK   WINDBREAK.

DAM

HEAVILY GRAZE OR SLASH AND REMOVE VEGETATION

ORCHARD

USE A HEDGE, STONE OR METAL FENCE AS A RADIATION SHIELD.

W TAP   LAWN

SWIMMING POOL.

PROVIDE GOOD ACCESS FOR EMERGENCY

ROADWAY

W TAP   TANK W

OPEN SIDED SHEDS SHOULD FACE EAST IF POSSIBLE

VEGETABLE GARDEN

W W TANK

YARDS

LOCATE SHEDS TO EAST AND SOUTH.

F F FUEL

PROVIDE LARGE OUTLET TO MATCH LOCAL FIRE BRIGADE EQUIPMENT.

ROADWAY FIREBREAK WINDBREAK WILL HELP TO INCREASE PROTECTION.

LOCATE FUEL AND HAYSTACKS TO SOUTH AND EAST OF HOUSE BUT KEEP SEPARATE.

HAYSTACK

Figure 5.2 *Windbreaks as a component of a low fire risk house block design. (Source: Garrett 1989.)*

Windbreaks create important plant diversity in the agro-ecosystem. This increases habitat opportunities for wildlife. Almost any windbreak will provide habitat for some species.

If improved wildlife habitat is a major reason for windbreak establishment, then the windbreak can be designed in a way to increase the number and diversity of wildlife supported. This can be done without compromising the other benefits that windbreaks provide. The major factors to consider are:
- the number and area of trees, shrubs and other plants
- the species of plants chosen
- the structural diversity of the windbreak, and
- the location of windbreaks to provide corridors linking larger habitat areas.

Broader multi-row windbreaks produce greater wildlife benefits than narrower windbreaks by simply providing a greater area of habitat. The addition of more rows to a windbreak can improve both its function to moderate the effects of the wind, and its value to wildlife. Wider windbreaks can incorporate a wider diversity of tree and shrub species than is simply required for shelter. Greater species diversity generally means more wildlife.

A wildlife corridor is a continuous strip of vegetation which links two larger areas of habitat and allows wildlife to migrate between them. Isolation of wildlife populations makes them much more vulnerable to local extinction, through limiting the area of habitat they can exploit. Impacts such as a fire can be much more severe on isolated populations. In the long term, wildlife can also decline through the effects of in-breeding.

The usefulness of wildlife corridors depends on the requirements of the species for habitat and migration. Obviously, wildlife species which are dependent on trees and shrubs will be particularly benefited by the establishment of windbreaks in otherwise sparsely treed areas, and through the linking of larger areas of vegetation with tree corridors. Other non-dependent species may also benefit to some extent through the provision of additional food, shelter and nesting sites. Adding blocks of vegetation along a thin corridor greatly increases its habitat value. Windbreaks can link with other linear areas of vegetation such as along roadsides and streams to create a web of vegetation in an area. A web of interconnected vegetation is much more likely to sustain significant wildlife populations in the long term than a collection of isolated remnants.

Flowering trees and shrubs provide food for nectar-feeding birds and also attract insects which provide food for numerous insectivorous birds and mammals. It is a good idea to achieve some flowering throughout the year by planting a diversity of species with different flowering times. Dense prickly shrubs are often excellent habitat for a wide variety of wildlife species.

The use of plants indigenous to an area will generally provide the best habitat requirements for the original wildlife. This is because this wildlife has co-evolved with and has adapted to these species. Indigenous plants also generally grow well because they are adapted to the local environmental conditions. The use of local indigenous plants helps preserve the genetic diversity of local species as well as maintaining unique local landscapes. Non-indigenous species can, however, still provide valuable wildlife habitat.

As well as large trees and shrubs, it is relatively simple to incorporate the planting of understorey plants (native grasses, herbs, ground cover, small shrubs) into a windbreak. This creates a much more diverse vegetation community and greater benefit to wildlife. Also allow the accumulation of leaf litter, logs and other debris where these will not pose a fire threat. This can provide habitat for a variety of birds, reptiles and small mammals.

Dead trees with hollows are important nesting sites for many forms of wildlife and should be retained. If trees are not old enough to have hollows, artificial nest boxes can be provided.

Windbreaks and other vegetation near streams, dams and other water bodies can be particularly attractive to a range of birds and other wildlife

A major benefit of wildlife on the farm is that many species of wildlife (particularly insectivorous birds) eat enormous quantities of problem insects. These especially include damaging pasture pests such as scarab beetles, cockchafers, crickets and grasshoppers. This pest control costs nothing and is ongoing, year in, year out. Windbreaks are very useful in this respect because of the long boundaries they provide between pasture and bird habitat.

For example, ibis feed on about 200 g of insects every day. Magpies are renowned as consumers of vast quantities of pest insects of pastures. Insectivorous bats may eat more than half of their body weight in insects in a night (Platt 1995).

Many Australian farmer groups have planned and coordinated the establishment of wildlife corridors across their properties to link distant areas of remnant vegetation or to create "green webs" of vegetation in their district. Windbreaks, erosion and salinity plantings, roadside vegetation, and woodlots can connect and reintegrate remnant patches of bush. These remnants themselves are vital wildlife refuges and should be fenced and managed (for example, pest plants and animals controlled, regeneration encouraged). Green webs will produce a healthier, more visually appealing and sustainably productive rural environment.

The encouragement of wildlife is generally very complementary to agriculture. In fact the integration of nature conservation with productive farming is a common feature of emerging sustainable agricultural systems.

Sustainable Agriculture

Numerous definitions of sustainable agriculture exist. It is an idea which assumes different meanings to different audiences. Most agree that sustainable agriculture is about ensuring that the capacity of the land to produce what we want from it is maintained over time. Sustainable agriculture requires the integration of economic, ecological, and social values and constraints into planning and decision-making. By doing this, options of what can be produced in the future from the land are preserved.

The original vegetation communities in Australia's agricultural lands generally have a high degree of stability, resilience and ability to recover from disturbance and stress. These communities consist of floristically and structurally diverse plant communities with a high proportion of perennial plants. Nutrient, water and energy requirements in these ecosystems are met entirely from natural sources. Agricultural ecosystems are obviously different, being more productive and reliant on artificial inputs. Nature can, however, provide some important design principles for improv-

Figure 5.3 *A windbreak on the Lyons' property near Balmoral in western Victoria. The windbreak forms part of a web of green corridors planted by the Dundas–Black Range Corridor Group which aims to re-connect the isolated remant vegetation of Mt Dundas with the Black Range.*

ing the sustainability of agricultural production. We can learn to incorporate some of those elements of the original vegetation communities which contribute to system stability, and complement agricultural production.

An increase in the cover of perennial plants and level of bio-diversity is an important aspect of increasing the stability of our farming systems. This helps maintain the functioning of stable ecosystem processes such as nutrient and water conservation and cycling. The incorporation of perennial woody vegetation such as windbreaks into simpler traditional agricultural systems is a good way of achieving this increased stability. Planting windbreaks into agricultural systems can create a replica of the original vegetation communities, with many of its characteristics of stability and resilience.

This chapter has made the case that the presence of windbreaks can be an important part of a strategy to conserve fundamental soil and nutrient resources. The maintenance of these prerequisites for agricultural production is an important decision-making constraint in sustainable agricultural systems.

Practices which enhance agricultural sustainability are more readily adopted if they can be demonstrated to be economically beneficial in the short term, and are technically sound. We have seen in Chapter 4 that windbreak establishment is both practical and a good financial investment for most farmers. As well as helping to increase agricultural profitability, the diversification of farm incomes through the production of timber and tree crops can increase the economic stability of the agricultural system as well as its ecological stability.

Many agronomic approaches to increasing agricultural productivity require high levels of finance and artificial inputs. In contrast, the increases in production and profitability from windbreaks are achieved with minimal inputs of finance and resources once the windbreaks are established. The improvements to agriculture are long term and continuous.

While windbreaks are an important tool to increase the sustainable productivity of farmland, they are not the total solution. The adoption of no single land use practice will achieve sustainability. A great diversity of systems and approaches is required. Windbreaks should be considered part of a package of necessary practices.

Hedges — A Hard Learnt English Lesson

For centuries the hedgerow was a key feature of the English landscape, acting as a boundary marker between fields and farms, giving shelter to crops and livestock as well as providing a home to numerous wildlife.

Post-war agricultural changes saw paddock sizes increase enormously as thousands of kilometres of hedgerows were torn up to make way for bigger machinery.

Throughout the 50s, 60s and 70s agriculture boomed, yields soared, and machines became even larger and small farms with their small paddocks became rare as larger farms merged. But with the greater yields and bigger machines came even greater soil erosion as wind and water took its toll. The rich English earth now in such vast fields, had little protection from the elements. It seemed that a more ancient wisdom had been overlooked – not only did hedges divide neighbour from neighbour, contain livestock and appeal to the eye, but they were key elements in preserving and maintaining soil. As the hedgerows disappeared, so did the soil as it blew or washed away.

It is only now that the true irony of the policy is being seen. The Ministry of Agriculture, having led the drive for larger fields and greater yields paid out millions of pounds of taxpayers' money over almost four decades to encourage and subsidise hedgerow removal. It is now offering grants for farmers to replant hedges, something many farmers are doing under the "Hedgerow Incentive Scheme" in a bid to prevent any more of their valuable top soil blowing or washing away.

It appears that for all the gains of agriculture in both England and Australia, a few fundamentals have been overlooked – soil management against the elements. Alley farming, hedgerows and windbreaks are key strategies in soil husbandry – something our forebears perhaps knew more about than we give them credit for?

Figure 5.4 *Wade Muggleton, Cambridge College of Agriculture and Horticulture.*

More fundamentally, a change in attitude towards decision making on a long-term rational basis is necessary at the farm level. At the regional and national level there needs to be acceptance that agricultural production operates within the constraints of an ecological system.

Energy Conservation

Just as windbreaks alter microclimate for agriculture, they also provide microclimate benefits which reduce the heating and cooling energy requirements for buildings.

Heat exchange between the inside and the outside of a house (generally through cracks around doors and windows and other openings) is a major cause of energy loss in homes. The amount of heat exchange can be strongly influenced by windbreaks near a house. In North America, the

Figure 5.5 *A well-sheltered homestead is a pleasant place to live.*

establishment of windbreaks near houses has been shown to reduce heating costs by up to 15–25% (Heisler and DeWalle 1988).

Heat loss by conduction and convection through building surfaces also occurs, although the effects of windbreaks on this is less significant.

Similarly, the costs of cooling houses in summer with air conditioning may be reduced by providing protection from hot winds with windbreaks. Care should be taken not to locate windbreaks which may block cooling summer winds. Trees which provide summer shade also significantly reduce home cooling energy requirements.

CHAPTER 6
Farm Planning

Windbreak Location for Maximum Benefit

Windbreaks and the Whole Farm Plan

It pays to give careful thought prior to establishing a shelter system to ensure that the optimum benefits are provided in the shortest period of time. With good planning followed by careful establishment and maintenance, significant benefits can be achieved in just a few years.

Long-term improvements and changes to a property such as fencing, water supply or trees need to be carefully considered to maximise their value. A windbreak is a long-term feature of the landscape. It will probably outlast the fence which protects it many times over. The siting of such a long-term asset deserves careful consideration.

Whole farm planning is the ideal way to ensure that this is done. A good farm plan invariably identifies numerous opportunities for windbreaks and other vegetation to improve your property.

Whole farm planning ensures that there is a long-term strategy to guide farm design and management decisions. Future activities are planned, taking into account all of the physical aspects of the farm (soil, topography, water, vegetation, existing improvements) and financial resources to meet the economic and personal objectives for the property. This means that the farm is viewed as a system, and all the ramifications of any decision are considered.

While agricultural systems vary considerably across Australia, whole farm planning is a process that can be applied to any region.

If you are not intending to directly prepare the whole farm plan, you ought to be closely involved so that you clearly understand the planning process and the reasons behind the decisions.

Whole farm planning generally involves the following steps:

Goal Setting

This involves reflection and sometimes soul searching. What am I trying to achieve with my farm? In what condition would I like to see it in 20, 40, 100 years' time? A sense of long-term vision is important for a good whole farm plan. Long-term planning is essential in developing sustainable farming systems.

Don't forget it is your plan, so keep it workable and relevant to your needs.

Farm Analysis

A mapping exercise is undertaken to provide the basic resource information for the whole farm plan. The base map for the plan is usually an aerial photograph of the property of suitable scale, but in some circumstances a topographic map is used. A series of plastic overlay sheets are then generally taped to the base map. There is considerable potential for the increased use of computer-based Geographic Information Systems in the development of whole farm plans.

Although the approach can vary, three overlays are frequently used:

Overlay 1 — Mark land features

Your local knowledge of your land will be the most valuable information here. Careful observation is the key. The land features may include:

- Property boundaries. Mark in features on neighbouring properties if these are relevant to you.
- Drainage lines and streams. These are often associated with remnant vegetation.
- Ridge lines.
- Slopes/escarpments.
- Swamps/wetlands. These sensitive areas require careful management.
- Remnant vegetation. A precious asset, particularly if it contains intact understorey. Once remnant vegetation is destroyed or damaged it can be difficult to rehabilitate, and can never truly be replaced. Think carefully before undertaking any activity that may harm remnants in

generally cleared landscapes. Several states in Australia have introduced vegetation clearance control legislation. Many land degradation problems are due to over-zealous removal of native vegetation.

- Land classes. Land classing divides your property into areas of similar capability, each having different management requirements. Classing systems and definitions vary from state to state. Find out if a standard set of classifications exist for your area. If these standard classifications exist for your region, make sure that they make sense to you.
- Land degradation features; for example, gullies, sheet erosion, tunnel erosion, saline areas, stock camps, weed and pest animal infestations etc.
- Areas with poor pasture/crop growth.
- Particularly exposed areas requiring shade or shelter.

Overlay 2 — Mark existing improvements
(overlays 1 and 2 can sometimes be combined)
- Sub-divisional fences, paddock names and sizes.
- Gates/grids.
- Buildings.
- Water supply (dams, troughs, pipelines, bores).
- Roads, tracks, laneways, easements, firebreaks.
- Underground/overhead utilities.
- Woodlots, windbreaks and other plantings.

Plan Synthesis

In this phase, try not to be influenced too much by the existing layout but try to "take off the blinkers". It can be a good idea to seek outside opinions to raise awareness of opportunities you may never think of yourself. This is a good time to use one of the computer models described in Chapter 4 to determine the economically optimal windbreak layout.

Good technical advice is important for management changes you are not familiar with. The final overlay of the whole farm plan is your future vision for the property.

Overlay 3 — Mark proposed improvements
- Soil conservation earthworks. These may include gully head structures and diversion banks. Usually these should be implemented in association with other catchment improvement works.
- Land to be withdrawn from production. Analysis can highlight how gross margins on some land classes can be marginal or negative.

- Woodlots, agroforests, windbreaks, stock havens, wildlife corridors, erosion and salinity recharge area plantations, and other plantings. Don't forget trees can provide multiple benefits. Several of these functions can be performed by the one planting, given careful layout and design.
- New fencing. Fences are expensive, so take care when locating them. Whole farm planning enables you to fence and subdivide your land according to its land class. Paddocks consisting of relatively homogenous land type can then be managed within their capability. It is often desirable to subdivide land classes into smaller paddocks. This can lead to more precise and efficient utilisation of pastures by stock. Odd corners and narrow areas created by re-fencing can be ideal locations for woodlots or windbreaks. Fencing of remnant vegetation is frequently necessary to ensure its long-term viability. Forward planning of fencing realignment adds little extra cost over time as it involves waiting until the old fences require replacement. High quality fences are essential for windbreaks and other revegetated areas as they will often be placed under greater pressure by stock than normal subdivisional fencing.
- Crop, pasture and livestock management proposals for each paddock. This may also be written down elsewhere in greater detail.
- Farm water supply improvements. This may include the establishment of new bores or dams and perhaps the installation of water reticulation systems to deliver water to troughs in paddocks. Watering points are areas of high stock concentration and therefore should be carefully located to minimise soil erosion and other damage.
- Tracks and laneways. Give careful consideration to access for vehicles, machinery and stock to paddocks. Improved access and capacity to move stock quickly is often a major benefit obtained through whole farm planning.
- Fire breaks. Think about the location of valuable assets and the likely fire approach direction when locating strategic firebreaks and windbreaks.

Scheduling Works

While the whole farm plan presents your long-term vision, it doesn't tell you how to get there. You need to schedule works according to a shorter time frame (say the next year) to put it all in perspective. This breaks down the challenge into more achievable chunks. List the tasks you want to implement in this time, their estimated cost, their priority and anticipated timing. Be realistic with what you can achieve with your financial and human resources.

The amount of time required to establish the windbreaks and other improvements identified in the plan depends on the scale of the works, and the financial and physical resources available. Take care not to be overly ambitious in the early stages. This can sometimes lead to disappointment due to poor establishment and maintenance. If you have little experience in tree planting, it is often a good idea to start on a small scale and build up confidence. Many farmers find it best to undertake an annual planting program for five or more years. Progress with this approach can be surprisingly rapid. It also spreads the risks between the seasons, as occurs in other farming practices. On an annual basis, list the tasks you want to implement, their budget, their priority and anticipated timing.

You may eventually aim to provide shelter over large areas or perhaps the entire property. While it is very desirable to take this long-term view, it makes sense to initially locate windbreaks where they will provide the highest benefits for the least cost. Try to determine the high priority areas requiring shelter. These may include around the homestead, lambing or calving areas where losses are consistently high, holding areas for off-shears sheep, or sensitive horticultural crops. Shelter can then be extended to other areas over time.

Review

A good whole farm plan is rarely set in concrete, but is reviewed and changed regularly. Planning is an on-going process so don't consign your plan to the bottom draw. Keep it somewhere where you will see it and use it regularly.

Siting Considerations for Windbreaks

Orientation

Take into account prevailing winds and critical winds and the likely locations of sensitive crops or stock. Note that prevailing winds and critical winds are often from different directions and require different placement of windbreaks.

Think about the winds which generally do the most damage, and locate windbreaks at right angles to these. This orientation will minimise the number of windbreaks required to provide a given level of protection across a paddock. Remember that the most destructive winds often come from a different direction to prevailing winds. In much of southern Australia it is the severe southerlies and south-westerlies which cause the greatest livestock losses.

Protection from hot drying winds in the late spring and summer may be particularly valuable for field crops and pasture production. This increases moisture availability to crops at this critical time, and reduces abrasion from wind-blown soil. Consequently, to maximise these benefits to plant growth, it makes sense to orientate some windbreaks in the farm plan at right angles to these winds. Drying winds are often northerlies and north-westerlies in southern Australia, so this means east-west running windbreaks. Research at the Rutherglen Research Institute in north-eastern Victoria has shown that east–west windbreaks were particularly beneficial to cereal crop yields, although increases were still significant for other orientations.

An inspection of meteorological records from nearby recording stations can provide valuable information about the distribution of wind speed and direction throughout the year.

It is rare that critical winds blow exclusively from one direction. A straight windbreak will only provide the optimum area of protection from winds generally at right angles to it. Small deviations from the perpendicular have little effect although, as the angle of the wind to the windbreak decreases, the area sheltered is reduced at an increasing rate. Some shelter is still provided, however, when winds are parallel to a windbreak (due to the boundary layer effect) and even on the windward side (refer to Chapter 1 for more detail).

In areas where strong winds can be experienced from many directions, windbreaks meeting at right angles maintain larger areas of shelter as the wind shifts. This is normally achieved by running windbreaks along paddock boundaries which meet in the corners.

Figure 6.1 *Plan view of the general pattern of significant reduction in windspeed near a windbreak for winds of different directions.*

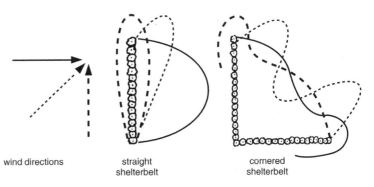

wind directions   straight shelterbelt   cornered shelterbelt

Establishment Costs

The cost of the establishment is another important factor when determining priorities for windbreak establishment. Economy of fencing, ease of establishment and adding value to existing shelter and other improvements are important considerations. Other revegetation options can also provide shelter benefits if well designed, such as fencing off and revegetating eroding gullies, or fencing off and regenerating remnant vegetation.

It is most common to locate windbreaks adjacent to existing fences. This certainly makes sense in terms of minimising fencing costs (generally the most expensive part of establishing a windbreak). Make sure, however, that existing fencelines do not encourage the establishment of a windbreak in a location where it simply isn't ideal. The value of whole farm planning is that it allows you to look beyond the current fencing pattern and picture the long-term optimal layout. Re-fencing paddocks into relatively homogenous land units is a common aim of whole farm plans. This enables areas of similar land type to be managed within its capacity to produce on a sustainable basis. This re-fencing provides an ideal opportunity for the incorporation into the farm layout of windbreaks along new paddock boundaries.

However, large paddocks may not necessarily be adequately sheltered simply by establishing field perimeter windbreaks. If shelter benefits are to be maximised, additional windbreaks need to be considered.

Special Windbreaks for Livestock Areas

In livestock areas with large paddocks, consideration should also be given to the establishment of carefully located, within-paddock shelter. Observe the movement of stock in

Figure 6.2 *Designs for within-paddock shelter developed by the Potter Farmland Plan project in western Victoria. (Burke et al 1988.)*

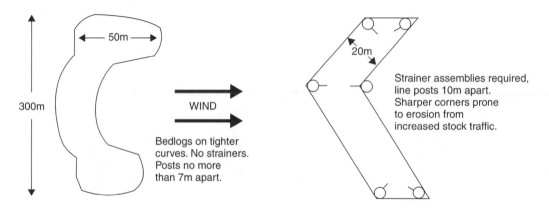

50m

300m

WIND

Bedlogs on tighter curves. No strainers. Posts no more than 7m apart.

20m

Strainer assemblies required, line posts 10m apart. Sharper corners prone to erosion from increased stock traffic.

Figure 6.3 *A recently established within-paddock boomerang windbreak. (Photo Andrew Campbell.)*

Figure 6.4 *A livestock haven is a small confined area of high quality shelter for sensitive stock under extreme conditions.*

bad weather. Sheep in particular will often end up in the downwind corner. The Potter Farmland Plan demonstration farms near Hamilton in western Victoria have successfully pioneered boomerang- and kidney-shaped windbreaks to protect these critical areas.

Livestock havens is another idea with considerable merit for exposed farms with no naturally sheltered areas for lambing or other sensitive livestock. This is essentially an area enclosed by windbreaks providing a very high degree of wind protection. Many well-timbered farms have still incurred high stock losses in severe weather. The key is to have areas of high quality shelter available and, importantly, people on hand to move stock at these critical times. Woodlots and agroforests can also be ideal areas to locate vulnerable stock in severe conditions.

Windbreaks are extremely useful when located along laneways, as stock can be easily held in lanes under extreme conditions. In high rainfall areas, be careful to locate windbreaks near laneways to avoid winter shading. Laneways can provide a very effective backbone for a shelter network on grazing farms.

Shade

The patterns of shade cast should also be considered when orienting windbreaks. East–west oriented windbreaks will cast the greatest area of shade in the middle of the day (usually when it is most needed) across the adjacent paddock. In a multi-row east–west windbreak, the area of shade cast will be maximised by placing the tallest row on the southernmost side.

North–south windbreaks will cast shade along the windbreak during the middle of the day. Where the effects of shade are to be minimised, for example with shade-sensitive crops, north–south windbreaks should be considered.

Other Factors

The location of windbreaks may also be affected by other considerations such as the provision of erosion control or wildlife corridors.

Spacing between Windbreaks

To maximise the benefits that windbreaks can provide, they need to be located more regularly than the randomly located and isolated windbreaks commonly seen. An infinite range of variation of this is possible, from "checkerboard" patterns of field perimeter windbreaks to more natural approaches with the location of windbreaks dictated by landform and the natural attributes of the land.

The ideal spacing between windbreaks will depend on windbreak design and the degree of shelter required.

Some protection is also provided in a narrow zone on the windward side of windbreaks. This is a factor commonly overlooked in windbreak siting and design.

The need for establishment of regular patterns of windbreaks has lead to the development of a range of windbreak systems which are described by the general term "alley farming". Alley farming is a farming system where agriculture is undertaken between a regular series of windbreaks. Alley farming systems are described in more detail in Chapter 7.

Topography and the Location of Windbreaks

Farmland is rarely totally flat. Topography changes the flow of wind and consequently affects considerations for the location of windbreaks.

Topography can provide naturally sheltered areas (for example, in gullies or valleys) which obviously benefit little from the establishment of windbreaks. Wind exposure gradually increases as you move up hillsides, with the upper slopes and hilltops generally being the most exposed. Planting contour windbreaks on the mid-slopes can protect these more exposed areas. Planting on the contour is desirable for most hilly terrain. Windbreaks along ridges are generally not recommended in hilly areas, as little additional leeward protection is provided. It is also more difficult to successfully establish trees on shallow ridgetop soils. In undulating country, windbreaks on the rises may be beneficial if they protect one or more downwind undulations.

The establishment of dense windbreaks on the contour on a hillside or across the end of a valley may trap draining cold air and induce localised frost. If this is a problem for sensitive crops, it can be minimised by the provision of gaps or ensuring the windbreak has reasonable porosity.

Windbreaks and Irrigation Farms

The development of intensive farming systems in irrigation areas frequently results in the loss of original trees with little replacement. Windbreaks can play an important role on irrigation farms in restoring some tree cover.

Windbreaks in irrigation areas often have spectacular growth rates because of the additional water available. They also have the benefit of not competing nearly as strongly for moisture with adjacent crops or pastures as non-irrigated windbreaks because the artificially applied water replaces water used by the trees. The reduction of evaporation near windbreaks also means that irrigation water is used more efficiently and less needs to be applied.

Irrigation salinity is largely caused by excessive recharge of saline groundwater from applied irrigation water. This can also be reduced by windbreaks and other tree establishment in combination with a range of improved irrigation practices. Irrigation seepage into the groundwater is decreased by windbreaks alongside irrigation water channels, drains, dams and irrigated paddocks.

Windbreaks can easily be incorporated into most irrigation layouts. Plan the location of windbreaks as part of any new levelling and layout work. As for dryland farms, windbreaks should be oriented at right angles to critical winds to ensure that the maximum area of protection to these will be provided.

Single row windbreaks are a common feature of irrigation areas, due to space limitations. These are often planted along irrigation bay check banks. Single row windbreaks should be composed of species which retain their foliage down to ground level. Check banks may need to be made slightly wider to accommodate the trees. Where space permits, wider multi-row windbreaks are recommended.

Other opportunities for windbreaks on irrigation farms also exist. Unproductive areas, especially wet areas where drainage water collects, or along water channels (providing channel access is not disrupted) are good sites. Windbreaks along channels can be useful in intercepting ground water seepage. Laneways are another good site for windbreaks. East–west laneways should be planted on the south side, and north–south windbreaks on the east side to minimise shading of the track.

Centre pivot systems create some special problems for windbreaks. Essentially two windbreak options exist for these. A circular windbreak or rectangular perimeter windbreak (perhaps with woodlots in the corner unused areas) can be established to provide perimeter shelter. This will not, however, be sufficient to provide optimum shelter benefits for large systems. The only option here is to establish additional circular low shrubby or herbaceous windbreaks whose height does not exceed the clearance of the irrigation equipment.

Locating Windbreaks Near Houses and Other Buildings

We have seen that carefully locating windbreaks around the homestead can reduce wind and dust, provide noise reduction and visual screening, protect the assets from fire, reduce heating and cooling costs, and provide a more pleasant environment to live in.

It is common to locate windbreaks so that the building is in the zone of maximum wind speed reduction. This varies with windbreak porosity (see Chapter 1) but generally occurs in the zone 2–5 H. The same basic planning principles apply as for agricultural land, namely, identify the most

Figure 6.5 *Well-sheltered irrigated pasture in northern Victoria, grazed by dairy cattle.*

critical winds and orientate windbreaks as close to perpendicular to these as possible. In areas with critical winds from a number of directions, two or more windbreaks at right angles may be necessary.

We have already seen that strategic location of windbreaks can provide fire protection benefits by reducing wind speed and thus the intensity of fire, catching or deflecting airborne material, absorbing radiant heat, and deflecting heat and smoke-laden winds. Buildings and open spaces in the lee of windbreaks are protected to various degrees from the effects of fire. The zone with the greatest wind speed reduction (3–5 H), where wind is usually reduced by more than 50%, provides the most protection. Note that it is hazardous to locate buildings immediately adjacent to a windbreak, especially where it may catch fire. Low flammability species are highly recommended for planting near buildings. In southern Australia, days of extreme fire danger are often characterised by north to north-west winds which may change to the south-west late in the day. In fire-prone areas, windbreaks near houses should be located bearing in mind likely fire directions.

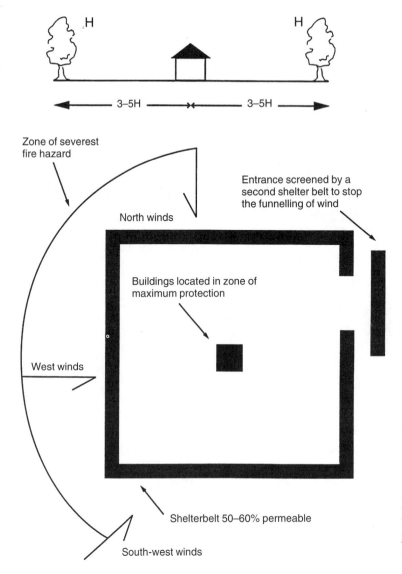

Figure 6.6 *Locating windbreaks near houses to provide maximum fire protection benefits. (Source: Country Fire Authority 1992.)*

Natural Shelter

Windbreaks will naturally provide the greatest benefits in regions with a high degree of exposure to wind. This includes huge tracts of the wheat/sheep belt in southern Australia. Many other areas of Australia have a degree of natural shelter due to topography and/or remnant tree cover and therefore the need for windbreaks is less widespread. For example, in the monsoonal region of Queensland, present levels of tree cover are thought to be generally adequate for providing shade and shelter to grazing animals.

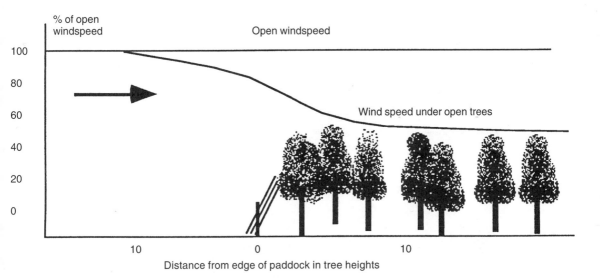

% of open windspeed

Open windspeed

Wind speed under open trees

Distance from edge of paddock in tree heights

Figure 6.7 *The reduction in wind speed by scattered mature trees. (Reid & Stewart 1996.)*

Scattered remnant trees through the landscape can also be effective in reducing wind speed. At Vasey in western Victoria, scattered river red gums (*Eucalyptus camaldulensis*) at 17 trees per hectare have been shown to reduce the wind speed to 50–60% of that in adjacent open paddock. This is the result of the trees producing an increase in the general surface roughness of the landscape rather than a simple conventional windbreak effect. Timber woodlots or agroforests can provide very valuable shelter for exactly the same reasons. Remember that the amount of feed available is likely to be limited underneath trees, so stock can generally only be left in these areas for a limited time.

The cost of re-establishing scattered trees is usually far greater and the results less successful than establishing windbreaks or clumps and cluster plantings of trees. The maintenance of existing remnant trees should therefore be given a high priority.

Valuable shelter can also be provided for lambing ewes and off-shears sheep by remnant areas of tussocky native grasses.

The arid and semi-arid areas of Australia are extremely prone to erosion. Only 10% of shrublands and 30% of grasslands remain in good condition. Grazing of natural rangeland vegetation is the major agricultural enterprise in these regions. Maintenance of good levels of natural vegetation cover, particularly during drought, through the adoption of appropriate grazing strategies, is the key to maintaining shelter and the sustainable agricultural productivity of these areas.

CHAPTER 7
Alley Farming

Alley Farming Basics

We have seen how windbreaks can provide a wide range of benefits to Australian farms. The effect of a few randomly located, isolated windbreaks will, however, be fairly small. To maximise the benefits of windbreaks, they have to be established in a regular pattern across the landscape. One approach to achieving this is alley farming. Alley farming is a relatively new farming system to Australia which involves growing crops or pastures in "alleys" between a regular series of parallel windbreaks.

The important principle in alley farming is that the tree and agricultural production can be complementary, producing more than either a tree or shrub plantation or conventional agriculture alone. There are two major reasons for this — first, the shelter benefits that windbreaks provide to crop and pasture growth and soil protection, and second, the mixing of trees and agricultural plants more effectively exploits the moisture, sunlight and nutrients available on the site. This is due to their complementary distribution of foliage and root systems, and the fact that their major growth periods often occur at different times of the year.

The trees and shrubs in the alley farming system, as well as providing shelter and land protection benefits, can also be selected to provide timber, tree crops and fodder.

For alley farming to be commercially viable, the economic benefits from increased agricultural production due to the provision of shelter and directly from the trees and shrubs must exceed the costs of windbreak establishment and losses due to the land taken up by the windbreaks and their competitive effects.

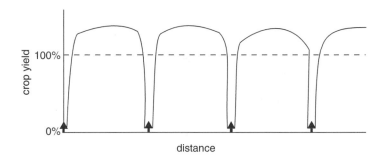

Figure 7.1 *The general effect of alley farming on crop yields.*

Numerous approaches to alley farming are possible on any site. Key design considerations are

- the kinds of crops or pastures to be grown in the alleys
- the species of trees and shrubs in the windbreaks
- the design of the windbreaks, and
- the spacing between windbreaks and their orientation.

Perhaps the most important of these design considerations is the spacing between windbreaks. A number of factors determine the optimal spacing between windbreaks to maximise agricultural production:

- the nature of the crop response created by the windbreaks
- the width of the windbreaks
- the width of the competitive zone beside the windbreak and the severity of the competition
- the height of the windbreak at maturity
- the lifespan of the windbreak, and
- the number of years the windbreak takes to reach its mature height.

The effect of spacing between windbreaks on crop productivity for just one possible combination of the above factors is shown in Fig 7.2.

For most pastures and field crops which are moderately responsive to shelter, recommendations for the ideal spacing for windbreaks is commonly in the range 10–25 H (tree heights). The aim here is to place the next windbreak where the yield increases due to shelter begin to decline significantly so that the shelter boost to crop growth is repeated. This spacing may be reduced for particularly sensitive crops or to provide protection to highly erodible soils.

Alley farming systems are particularly suited to highly mechanised cropping systems where the alley width can be

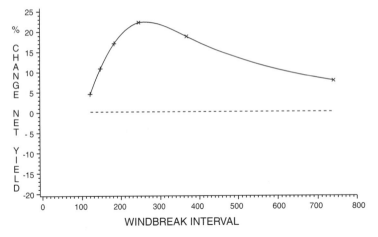

Figure 7.2 *The effect of windbreak spacing based on net oat yield responses to shelter at the Rutherglen Research Institute. This assumes the large oat yield response to shelter at Rutherglen shown in Fig 2.2, windbreak width = 10 m, windbreak height at maturity = 20 m, lifespan = 70 years, years to maturity = 30 years.*

designed around multiples of the width of commonly used machinery. Another benefit is that headland areas created by corners are minimised.

The optimal windbreak spacing will be influenced by whether or not yield responses are experienced on both sides of a windbreak. This is a common observation from research where yields on both sides have been measured. On sites where the important winds come predominantly from one direction, the response on one side of the windbreak may be greater than the other. In this case, the windbreak spacing may need to be reduced.

Alley farming varies greatly from region to region. The ideal spacing between windbreaks also depends on the objectives of the system. In grazing areas, alley farming may have a layout similar to cropping areas or the alleys may be smaller to create livestock havens of higher quality shelter to protect stock in severe cold and wet weather. In livestock areas, alley farming can also be partially or completely comprised of more closely spaced windbreaks of fodder shrubs such as saltbush, tagasaste and *Acacia saligna* which are either grazed or cut for fodder. Where timber production is particularly lucrative, economic production can be maximised by reducing spacing between windbreaks and increasing their width to produce more tree crops at the expense of agricultural crops.

The width of the competitive zone has a particularly important effect on the overall level of production from alley farming. This is because it can vary widely and occurs on both

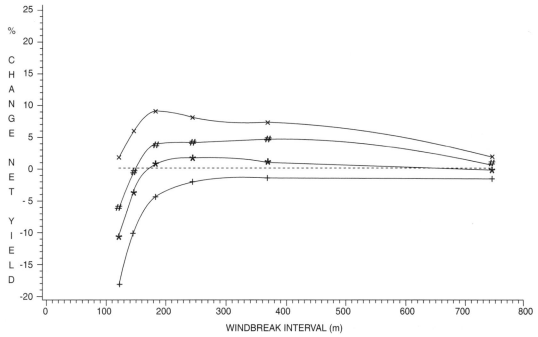

Figure 7.3 *The effect of windbreak spacing based on wheat yield at the Rutherglen Research Institute showing effect of two different levels of wheat yield response to shelter and two levels of extent of the competitive zone.*
+ = small positive wheat yield response to shelter and 2 H competitive zone,
* = small positive wheat yield response and 1 H competitive zone,
 # = moderate positive wheat yield response, x = moderate positive wheat yield response and 1 H competitive zone.

sides of windbreaks in an alley farming system. A wider competitive zone greatly reduces overall agricultural production. The root distribution of trees and shrubs largely determines the extent of competition for soil moisture. Some desirable tree and shrub species have deep roots which compete little with adjacent shallow-rooted crops and pastures. Other species produce very vigorous lateral roots which can extend several times the windbreak height into the adjacent field. Good establishment of trees and shrubs along deep ripped lines also encourages deep rooting. Where roots appear to extend into the field, regular deep ripping along the edges of windbreaks severs lateral roots and minimises competitive effects. Competitive effects due to shading can also occur. North–south oriented windbreaks cast the least shade onto the adjacent field and east–west windbreaks the most.

Alley farming is not widely practised in tropical and sub-tropical northern Australia, despite the fact that alley farming systems were originally developed in tropical developing countries. Here the major aim of alley farming is to conserve and improve soil on marginal sloping tropical croplands. This frequently involves relatively narrow alleys 10 m wide or less between hedges of multi-purpose leguminous shrubs. These hedges are regularly pruned for either fodder or green manure as well as for fuelwood. Pruning also minimises shading of crops (Kang et al 1993). More recently, increasing concern has been expressed about the adverse effects of close-spaced windbreaks/hedges on the productivity of tropical crops (Ong 1993) due to excessive root competition.

Alley farming is an important tool in the development of more sustainable cropping systems in Australia's fragile soils. As well as production from tree crops and shelter benefits to crop production, alley farming systems help reduce wind erosion and groundwater recharge, increase nutrient cycling and reduce artificial inputs. Other environmental benefits such as flora and fauna conservation can also be provided. The introduction of a continuous distribution of perennial woody plants through alley farming has the potential to increase the resilience and productivity of agricultural areas. The regular spacing of windbreaks across the landscape achieves this more effectively than the establishment of random and isolated blocks of deep-rooted vegetation. Alley farming creates an analogue of the original stable vegetation structure with a mixing of deep-rooted perennials with shallow-rooted annual species.

Figure 7.4 *Don Haines – Gerang Gerung, Wimmera, Victoria.*

Durable Hardwoods in Cropping Alley Farming

Don Haines farms 500 ha of mixed livestock and cropping at Gerang Gerung in Victoria's Wimmera. He recently embarked on the major challenge of establishing an alley farming layout over the entire property. Parallel windbreaks are being established 150 m apart, oriented SSE–NNW in conjunction with the establishment of a rotational grazing system. Windbreaks are protected by low cost electric fencing. The windbreaks will provide shelter benefits to crops, pastures through improved moisture availability, and livestock performance.

Windbreaks are native timberbelts consisting of slow-growing durable hardwood species including red gum (*Eucalyptus camaldulensis*), sugar gum (*Eucalyptus cladocalyx*), spotted gum (*Eucalyptus maculata*), red ironbark (*Eucalyptus sideroxylon*), drooping she-oak (*Allocasuarina verticillata*), and river she-oak (*Casuarina cunninghamiana*). Don believes there is great potential for these species to produce high value decorative and furniture timbers on a 50 to 60 year rotation.

While this is a long-term proposition, Don feels this is more like the management planning horizons that need to be taken to develop sustainable farming systems according to the long-term land capability. These thoughts were provoked by a trip he took to Europe visiting a farm which had been in the same family for 13 generations.

by Ron Dodds

Some Emerging Australian Alley Farming Systems

Timberbelt/Field Crop Alley Farming

Timberbelt/field crop alley farming is currently being pioneered in parts of Victoria, South Australia and Western Australia.

Windbreaks in this system consist of rows or a row of timber species generally grown in combination with complementary low-shelter shrubs in a typical timberbelt formation. Good financial returns from trees in the timber rows should be possible if they are carefully established and pruned to produce high value clearwood timber (more information on pruning trees for timber is provided in Chapter 10).

Selection of deep-rooting non-competitive tree species is important, particularly in drier cropping areas, sensitive to drought stress. Commonly, slower growing durable hardwood timber species are selected in drier cropping areas to produce a long-term high value timber crop. In wetter cropping areas, faster growing timber species could be considered.

If sheep or other livestock are to be used to graze stubble or pasture in a pasture phase, then fencing of the belts may be necessary. Low cost electric fencing should be considered, as permanent, non-electric fencing would generally prove to be cost prohibitive. If few or no livestock will be grazed on the paddock, then of course fencing may be unnecessary. Fencing may also be eliminated if stock are excluded from the paddock until the trees are above browse height or if less palatable tree species are selected. Careful grazing management can also be employed to ensure that stock are removed before damage to trees occurs. Sheep will tend to graze the pasture and stubble first before resorting to the trees.

Spacing between windbreaks is usually designed to optimise crop yield and minimise wind erosion. Frequently this means alley widths of 15–25 H, a considerable distance for tall windbreaks.

Windbreak orientation should be to maximise crop response. There is emerging evidence from the cropping regions of southern Australia that east–west oriented windbreaks providing protection from drying northerly winds are particularly effective. Crops are enhanced due to increased moisture availability, particularly at sensitive stages of crop development, and reduced physical damage.

Significant wind erosion control is also provided. In some severely erosion prone areas, particularly those with sandy soils, alley farming enables a crop to be sown and harvested where sand blasting and soil movement would normally make it difficult or impossible, even with minimum tillage cropping.

Another benefit of timberbelt/cropping alley farming systems is that they allow more flexibility in crop spraying and crop weed control. The lower erosion risk in alley farming systems may reduce the reliance on herbicides for weed control, and enable a greater amount of judicious cultivation. Lower wind speed means that spray drift is reduced and allows more flexibility with the timing of spraying. Care needs to be taken against spray drift into the windbreaks when using herbicides which may harm the trees.

Fodder Shrub/Livestock Alley Farming

For this particular application, windbreaks consist partially or entirely of fodder species, with pasture being grown in the alleys which are grazed by livestock.

Fodder shrubs commonly used include saltbush, tagasaste and *Acacia saligna* in southern Australia, and leuceana in the tropical north. These fodder species produce high yields of feed. The quality of this feed varies but can also be excellent (see table below).

Species	Nitrogen (%)	Digestibility (%)
Acacia saligna		50
Bluebush (*Maireana pyramidata*)	2.1	58
Leucaena	2.5	60–70
Mulga (*Acacia aneura*)	2.0–2.5	45–50
Saltbush (*Atriplex amnicola*)	1.5–2.0	55–60
Saltbush (*Atriplex nummularia*)	3.0	65–70
Tagasaste	3.6–4.6	72–77
Wilga	2.4	60
Willow	2.6	66

Source: Burke and Wilson (1997)

Most research into livestock production from fodder shrubs has shown that livestock performance is better when fodder shrub foliage is used as a supplement to conventional pasture. This is because fodder from shrubs alone may not always be of sufficient quality at key times, such as after lambing or fattening stock prior to sale. A common observation for stock feeding entirely on fodder shrubs is that they maintain condition but do not gain it. Fodder shrub/livestock alley farming systems consequently provide a desirable mix of pasture and fodder shrub feed.

The width of the fodder belts and their spacing depends on the proportion of fodder shrub feed to pasture feed desired. Commonly about 10–20% of the paddock area is occupied by fodder shrubs, which generally means a narrower alley width than for timberbelt/field crop alley farming systems.

Fodder shrubs utilise soil moisture deep from the soil profile which is unavailable to shallow-rooted plants and grow virtually throughout the year. They produce a considerably greater quantity of feed than conventional pastures. Fodder shrubs therefore are green and succulent when annual pastures are dry in summer. They also provide green feed in the critical late autumn/winter feed gap when pasture feed is scarce. These factors combine to enable a much higher carrying capacity than would be possible with pasture alone. Up to a four-fold increase in carrying capacity of some dryland farming land is possible through the establishment of fodder shrubs.

Fodder shrubs also minimise the requirement to conserve or buy in feed for critical times of low availability. They act as living haystacks and can provide important reserves of emergency drought fodder. This form of alley farming is particularly well suited to improve quantity and reliability of feed availability and overall carrying capacity for sheep grazing in the drier areas of Australia. As feed is available at times when it is otherwise scarce from traditional pastures, stock can be sold out of season in better condition and can achieve higher prices.

Pasture production in the alleys is enhanced by shelter from the fodder rows. Clearly, however, fodder shrub alley farming involves a compromise between the value of the fodder shrubs for shelter and their value for stock fodder. Two approaches are possible for the management of the shrubs for fodder. First, the fodder shrubs can be managed to maximise their fodder value, and as a result of this, reduce shelter benefits. This requires careful management and is described in more detail below. Alternatively, the fodder belts can be allowed to grow taller than browse height and be cut by hand or machine as an emergency measure in times of drought. This approach provides greater shelter benefits as the fodder shrubs are allowed to reach a much greater height. The value of the shrubs as a source of fodder is, however, reduced.

To maximise the feed value of fodder shrubs, careful attention must be paid to the grazing management of the alley farming paddocks. A cautious approach is required to grazing while the shrubs are establishing. Only limited grazing should be permitted in the first two to three years to enable the structure of the shrubs to develop. Gradually increase the intensity of grazing as the shrubs mature, and become more established.

Fodder shrubs respond best to short periods of intense rotational grazing. Continuous grazing is generally very damaging as the shrubs are not given the opportunity to regenerate, and foliage production is poor. Stock should be removed before excessive damage is caused to the woody structure of the shrub, particularly with cattle. Shrubs should be grazed frequently enough to prevent the development of woody stems unpalatable to stock. Sheep will eat stems up to 3 mm in diameter and cattle up to 7 mm. The ideal result is the creation of a lush, dense, multi-stemmed bush. Optimal forage yield is obtained by crash grazing two to four times a year (or three to six months apart) and allowing foliage regrowth in the intervening periods. The shrubs can generally be grazed down until there are still some leaves left and then stock should be removed. If the shrubs are grazed harder than this, the recovery period is much longer and some shrubs can die.

The long interval between grazing is far greater than is recommended for most pastures. This can lead to either undergrazing of pasture or if pasture is grazed optimally, then there can be overgrazing of the fodder shrubs. Temporary protection of hedges with electric fences for some grazing periods, or conserving some of the pasture grown as hay or silage can assist with these problems.

Research by Dr Laurence Snook in Western Australia has also shown that tagasaste and probably other browse shrubs have their palatability and feed value enhanced by fertilisation.

Generally the aim is to keep the fodder rows low, bushy, and at browse height of the stock. For sheep this is below about 0.8 m. If shrubs are ever allowed to grow above browse height, this can be corrected with pruning, forage harvesting, or slashing of the shrub rows.

Alley farming paddocks can provide important shelter benefits to livestock. In particular they may be used as livestock havens to prevent death of sensitive stock from exposure. If this is the case, fodder belts should be oriented at right angles to these critical winds.

Fodder Shrub/Field Crop Alley Farming

This is similar to the fodder/livestock system with the exception that alleys are cropped and occasionally grazed by livestock. Stock are introduced either to graze stubble after harvesting crops or pasture during a pasture phase. The fodder rows should have minimal competitive effects on adjacent crops. They will provide valuable fodder when livestock are introduced, combined with shelter benefits to crops and livestock.

This form of alley farming is particularly appropriate for mixed cropping/livestock farms where sheep production comprises an important part of the enterprise.

The width of the alleys should be similar to timberbelt/cropping systems (about 15–25 H), although measured in metres, this is obviously considerably narrower. Because the fodder trees are quite low, it may be necessary to space the fodder belts more widely than this optimum. Take care to ensure that the width of the alleys is designed to be a multiple of the pass width of the cropping machinery.

Figure 7.5 *An extensive fodder shrub alley farming system on the property of Dean Melvin at Dowerin, Western Australia. The carrying capacity of this land during the pasture phase has been increased approximately three-fold since the establishment of the fodder shrub belts.*

Fodder shrub/field crop alley farming requires almost no form of fencing — a major attraction.

This system was first used extensively by Western Australian sandplain farmer Dean Melvin who commenced establishing his alley farming layout in 1990. Fodder belts consisted of the leguminous shrubs tagasaste and *Acacia saligna* were established on his poorest sandy soil to reduce wind erosion and provide late autumn and winter feed. As a result, the need for additional feeding of stock has been virtually eliminated. Dean initially established windbreaks to produce alleys 30 m wide, but later increased this to 60 m. The result is that some of Dean's worst, least productive land has been transformed into a major asset.

Timberbelt/Livestock Alley Farming

Timberbelt/livestock alley farming consists of livestock production grazing on pasture between a regular series of timberbelts, and was originally developed in New Zealand, where it was established in conjunction with sheep grazing. It is being increasingly adopted in the higher rainfall pastoral regions of southern Australia.

The fact that the land is not cropped enables a greater variety of layout options than is generally achieved in cropping-based alley farming systems.

Tree growth in these higher rainfall areas can be very high, producing significant amounts of valuable timber. Returns from this, combined with enhanced agricultural production from the sheltered alleys can make timberbelt/livestock alley farming a very economically attractive option.

Damage of trees by stock is commonly experienced in this system where the trees are not protected by some form of protective fencing. Ultra low cost, low tension electric fencing systems have been developed to achieve this. An alternative is to cut hay in the alleys for four to five years until the trees are able to withstand grazing. This approach can lead to a decline in pasture quality, particularly perennial pastures and subterranean clover.

The windbreaks are commonly two-row timberbelts. The classic New Zealand timberbelt consists of a high pruned radiata pine row planted closely alongside (sometimes as close as 1 m) a row of low-shelter species such as *Cedrus atlantica*. As well as pruning for timber production, in New Zealand these windbreaks are also often side trimmed to minimise space occupied and shading of pasture. Numerous other species combinations are possible, including the use of native Australian timber trees. Where the aim is to produce greater amounts of timber from the site, the windbreaks can be widened to include a greater number of timber rows.

The timberbelts obviously provide valuable shelter benefits to both pasture and livestock. They should be oriented to provide protection from the prevailing and critical winds for livestock production and pasture growth. In cooler, wetter livestock areas, where shading of pasture can be more detrimental, north–south oriented windbreaks which minimise shade cast, as well as provide protection from cool westerly winds, may be desirable.

Ideal timberbelt spacing in this system would again normally be in the 15–25 H range. This may be reduced in some systems, for example where higher quality shelter for sensitive livestock is needed, such as for lambing ewes. Windbreak spacing may also be reduced where a greater volume of timber is to be grown on the site at the cost of reduced agricultural returns.

CHAPTER 8
Windbreak Design

Fundamentals of Good Design

A drive through the countryside unfortunately often reveals many poorly designed and haphazardly established windbreaks. Such faults are difficult to correct after planting. A little thought and care in the design of a windbreak will result in a valuable long-term asset and will improve a farm's productivity and profitability. The following are the key elements to think about when designing windbreaks.

Height

Maximising windbreak height is generally the most important windbreak design consideration.

The area protected by a windbreak is directly proportional to its height — doubling the height of a windbreak doubles the area it protects. Increasing the average windbreak height in a windbreak system increases the distance required between the windbreaks to provide the same level of protection.

Windbreaks should therefore be designed and established to make sure they reach the maximum mature height possible. The most important design issue is to include at least one row of the tallest shelter trees capable of growing on the site. Also, use good quality plants of species and provenances (local varieties) known to grow well in your locality. Provide good conditions for plant growth by following recommended site preparation and establishment procedures.

As well as the final mature height, other factors to consider when selecting the species in the tallest rows are their growth rate and the lifespan. Fast growth to the maximum height means the maximum extent of shelter is achieved earlier. Long-lived species increase the time span that this is maintained.

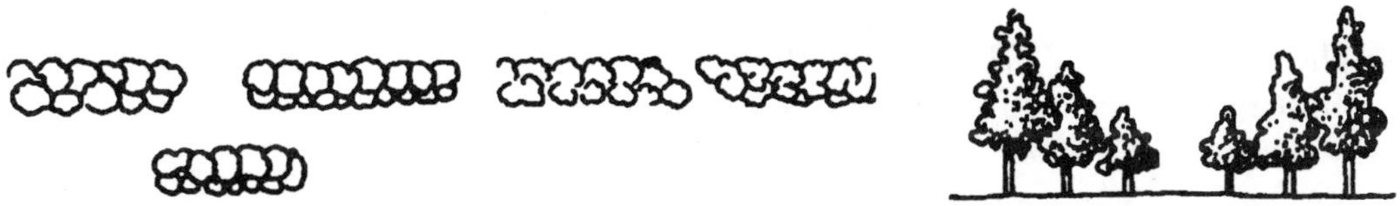

Island shelterbelt
(plan view)

Angled gap
(plan view)

Tapered gap
(side view)

Figure 8.1 *Some variations in windbreak gap design.*

Length

The length of a windbreak and the presence of gaps also greatly influences windbreak performance.

Again, big is beautiful. Short windbreaks tend to be very ineffective because the wind tends to curl around each end due to end turbulence, resulting in a relatively little area being protected. Try to make every windbreak as long as possible. A common rule of thumb is to make windbreaks at least 10 times as long as their mature height.

Gaps in windbreaks should be avoided if at all possible, as they decrease windbreak effectiveness. Windspeed in front of a gap is increased due to concentration and funnelling of winds.

Gaps are, however, sometimes necessary due to the need for gates and access through windbreaks. If this is the case, wind funnelling problems can be kept to a minimum. A small island shelterbelt just in front of the gap is one solution. In multi-row windbreaks, the gap can be angled at 45 degrees to the prevailing wind direction.

Porosity

Porosity is the degree to which wind can pass through a windbreak. How porosity affects the degree and extent of shelter is described in Chapter 1. Give careful thought to the pattern of wind speed reduction that you require behind a windbreak, then design it to have the porosity which will produce this.

In Chapter 1 we saw that the extent of significant wind speed reductions (say more than 20–30%) is not greatly affected by porosity for medium dense windbreaks (about 50% porosity) through to dense windbreaks (down to 0% porosity).

As porosity increases from 0% up to about 20%, the area with low wind speed (say, more than a 50% reduction) increases, and decreases thereafter. Windbreaks with more than 80% porosity tend not to reduce wind speed sufficiently to produce useful benefits, and are prone to problems with gaps. The usual aim for broadacre paddock shelter is to have uniform moderately dense windbreaks with optical porosity of around 20–40%.

There is a strong relationship between windbreak porosity and the minimum wind speed produced. Denser (or less porous) windbreaks reduce wind speed to a lower level immediately in their lee, and the point where the lowest wind speed occurs is located closer to the windbreak. Very dense windbreaks (< 20% optical porosity) are desirable where a small area of high quality shelter is required, such as livestock havens or around houses.

A significant problem with these very dense windbreaks is the greater turbulence they create in their lee. Turbulence generally decreases with increased porosity. This increased turbulence is felt at ground level beyond a quiet zone of reduced turbulence which extends to about 8 H. Hence significant wind turbulence can be expected in the zone extending beyond 8 H for dense windbreaks.

The vertical distribution of porosity is also important. In general, the usual aim is to have uniform porosity from ground level to the top of the windbreak. The most common mistake is to not provide low shelter down to ground level, either due to poor windbreak design or maintenance. The effect of this is funnelling of winds underneath the windbreak creating a narrow zone of increased wind speed immediately in the lee. Clearly this zone has the potential for adverse effects on crops, livestock and exposed soil. Beyond this, these windbreaks still provide an extended area of valuable shelter.

The type of trees in the windbreak greatly affects its porosity. Different species have greatly different foliage density and therefore produce windbreaks of different porosity. For example, eucalypts tend to have more open crowns while introduced cypresses have very dense foliage. Windbreak porosity can be reduced by reducing the spacing between trees within rows and by increasing the number of rows.

Number and Spacing of Rows

Single row windbreaks can provide very good shelter, providing they consist of tree species which retain foliage down to ground level. These windbreaks use the least area, but they are susceptible to problems caused by gaps and non-uniform growth. The maximum height obtained is frequently less than for multiple row windbreaks. These problems can be overcome by increasing the number of rows at the small expense of more space used. Given the significant investment in establishing and double fencing a windbreak, compromising on the number of rows is often a false economy.

It is usually a good idea to think about having multiple row windbreaks. Belts with several rows give a better opportunity to use tall tree species which naturally lose their foliage near ground level (or are pruned for timber production) in conjunction with lower shrubby species. Good permeability and height can often be obtained from two to four rows with one or two rows of tall trees. Narrow windbreaks of two to four rows are generally optimal in terms of minimising the area of land removed from agriculture while providing maximum shelter benefits.

Other benefits can be provided by further increasing the number of rows. Rows of fast growing species can be included to provide rapid shelter which will later either die out or be removed. Wildlife habitat is usually enhanced. Stock can be moved into fenced, broad, multi-row windbreaks at critical times when maximum shelter is needed. Tree and shrub species which are more susceptible to the wind and may be inappropriate in narrow windbreaks can, however, be included in the protected inner rows of a broad multi-row windbreak. In very exposed climates, such as in coastal areas, a larger number of rows can assist tree survival and growth due to greater mutual support.

More rows in windbreaks are particularly recommended for dry regions, because growth rates are generally slow and survival rates more variable than in wetter areas.

There is some evidence that very broad windbreaks produce an aerodynamically less efficient barrier to the wind than narrower ones. This should be considered before unnecessarily including a very large number of rows in a windbreak.

Valuable timber can also be grown if one or more rows of timber-producing tree species are included. In New Zealand, a system of timber-producing windbreaks called "timberbelts" has been developed, and is being widely adopted in southern Australia. Here a row or rows of high-pruned, timber-producing species is planted closely alongside a row of low-growing shelter species. More timber rows can be included if greater timber production is desired at the expense of less land being available for crops or pastures.

Rows should be about two to four metres apart to allow adequate room for the development of the trees and shrubs. In general, the fewer the rows, the closer they may be spaced.

There should be two and preferably three metres between the outside row and the fence to prevent stock browsing off too much low level foliage. Loss of low foliage results in undesirable increases in wind speed close to the windbreak due to wind funnelling. This distance can be reduced slightly with electric fencing or where the outside rows are to be trimmed. Leaving an adequate distance between the outside row and the fence leaves space for the trees to grow, and greatly reduces stock pressure on the fence and the likelihood that it will fail. An extra metre either side of a windbreak only occupies 0.2 ha per km of windbreak, a small price to pay for significantly enhanced windbreak performance.

Rows of low-growing trees and shrubs should be located on the outside of the windbreak to prevent them being shaded out by taller species.

Cross-sectional Profile

A common misconception in windbreak design is to aim for a sloping cross-section or profile. A sloping aerodynamic profile actually reduces windbreak effectiveness. Steep-sided windbreaks create a more effective barrier to wind flow and therefore shelter a greater area. Multi-row windbreaks should, if possible, have their tall rows on the windward side and the lower growing rows to leeward.

If shade is desired from the windbreak, then place the tallest rows so that they cast the maximum shadow at times when shade is most needed (usually mid afternoon).

Spacing of Trees and Shrubs within Rows

Spacing should be sufficient for the windbreak to provide the required porosity in a reasonable amount of time. Too great a spacing leads to an excessive period of time before the gaps close between the trees. Too small a spacing can lead to insufficient porosity and results in higher establishment costs.

Tree spacing should be varied according to the types of tree being used. The average spacing for medium to tall trees is 3–6 m, and for small trees and large shrubs 2.5–4 m. Compact low shrubs can be planted down to 1.5 m apart, but 2.0–3.5 m is more common for low-growing species. In general, trees and shrubs should be more closely spaced in narrower windbreaks and can also be closer in wetter, more fertile areas.

An alternative approach is to plant trees and shrubs more closely than the final intended spacing, and to thin them as they grow. This results in the faster establishment of shelter and can be achieved in a number of ways. The simplest is to plant the long-term shelter species at a greater density and thin them as they develop. Another method is to alternate the planting of fast-growing short-lived shrubs with slower-growing long-lived species. The faster-growing shrubs are thinned before they compete excessively with the long-term species which will ultimately comprise the windbreak. The trees and shrubs removed can be used for firewood, fence posts or fodder.

Try to stagger the position of trees so that they are opposite spaces in adjoining rows to provide more uniform porosity, fewer gaps, faster establishment of shelter and optimal use of the space available.

Tree spacings in some commonly used windbreak designs are shown in Fig 8.2.

Species

Plant tree and shrub species which will provide the windbreak characteristics you require (particularly final height, growth rate and porosity).

The most important considerations when selecting species are:
- At least one row of the tallest species that will grow on the site should be included as previously described.
- Foliage density and crown shape. This varies greatly with species, so affecting windbreak porosity. For single row windbreaks, the species selected must provide adequate foliage density from treetop to ground level. More flexibility in species selection is possible where multi-rowed windbreaks are to be established. Use of smaller trees and shrubs in particular can reduce problems with the loss of ground level foliage of the tallest trees.
- Growth rate. Fast early growth is desirable as it leads to the more rapid establishment of shelter. Unfortunately it is common for rapid-growing species to be short-lived. Trees which grow very fast can also be less wind firm. In these cases, a row of these species can be planted on one side of the windbreak or planted alternately with slower

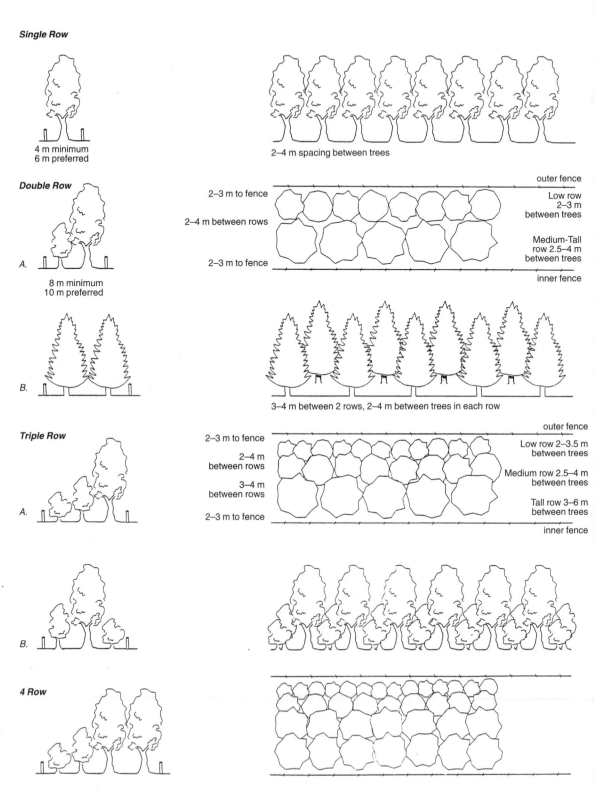

Figure 8.2 *Windbreak designs – tree spacings.*

growing species within a windbreak row. They can then be allowed to die out or be thinned when they are not required. Fast growth is also advantageous where the species are selected to provide timber, fuelwood or fodder production.

- Compatibility with crops and pastures. Some species compete with adjacent crop and pasture growth to a greater extent than others. Trees and shrubs with deep, penetrating roots generally compete less for moisture than species with shallow-spreading root systems. Some species chemically inhibit the growth of plants around them through a process called allelopathy. Conversely, other trees and shrubs can have beneficial effects, for example through nitrogen fixing (particularly allocasuarinas, casuarinas, and acacias) and adding organic matter to soil.
- Ability to regenerate naturally under the environmental conditions of the site. Choosing these species creates perpetual, self-sustaining windbreaks. Indigenous species are often able to regenerate most easily.
- The lifespan of tree species. Selecting long-lived species obviously maximises the useful life of the windbreak. Try to choose species which can live together harmoniously for a long time. By selecting species of roughly equivalent lifespan, impaired windbreak performance as plants gradually die out can be avoided. Be careful of the quickest growing species as these are often short lived or prone to disease. The benefit of quick shelter may be gained at the expense of a much shorter period of effective shelter. Alternatively, closely planted nurse rows of rapid-growing, short-lived shrubs or trees may be worth considering to provide early shelter for slower-growing species, especially on exposed sites. The nurse rows can be removed or be allowed to die out as the windbreak matures.
- Number of species in each row. Try not to include too many, particularly for windbreaks with only a few rows, as this tends to decrease uniformity in height and permeability. Often one species per row is best. There are some exceptions to this rule, such as the alternate planting of slow- and fast-growing species within a row (described above), or the alternate planting of tall pruned timber species of tree with slow-growing, shrubby shelter species within the same row.
- Fire resistance or ability to regenerate is desirable in fire-prone areas. This can save much in re-establishment costs.

For example, many eucalypts reshoot from branches and trunks after being burnt. Other species respond to being burnt by fire through massive germination of seed accumulated in the soil. This is common with wattles. The result is rapid regrowth of the windbreak at no cost.

- Flammability of foliage. Tree and shrub species of low flammability can protect assets against the heat of fires when used as radiation shields. These should be located well away from buildings as they will still burn under extreme fire conditions.

- Landscape and wildlife. Select species which blend in with existing vegetation and landscapes. Species can also be selected to provide food and shelter for wildlife. Indigenous species (those naturally occurring in your locality) usually offer the best value for providing wildlife habitat. They are part of the local environment, and local wildlife are well adapted to them.

- Timber and tree crops. Species can be selected so that timber and tree crops are produced directly from the windbreak. Timber can be grown in farm windbreaks without compromising shelter benefits. Firewood, posts and poles can be provided for on-farm use. These products, together with sawlogs and pulpwood, can be produced for sale to supplement and diversify farm income. To grow useable timber, care must be taken with windbreak design, and management, as well as species selection. Trees in windbreaks can also be selected to provide other valuable tree crops such as seed, nuts, honey and broombush.

- Coppicing ability. Trees which coppice (or re-sprout from cut stumps) reduce re-establishment costs after the trees are harvested for timber or fodder. Coppicing species can greatly assist with windbreak re-establishment and renovation.

- Fodder. Low windbreaks can consist entirely of fodder species of shrubs (for example, tagasaste, saltbush or *Acacia saligna*). In some alley farming systems, low fodder windbreaks are left unfenced and directly grazed by stock. Alternatively, a fodder row or rows can be included in a more conventional windbreak. If these are located close to the fenceline, the shrubs can be simply lopped and thrown over the fence to stock.

Most importantly, the windbreak species must be able to grow well on your site. Two rules of thumb can be applied here. Look around your district and see what species are healthy and being successfully used in windbreaks. Secondly, plant species indigenous to your locality. They are well adapted to the local environmental conditions, having evolved there over thousands of years. Seek advice from an experienced treegrower, nurseryman or government advisory officer about the windbreak species that have performed well in your area.

CHAPTER 9
Windbreak Establishment

Tree Planting

Weed Control

Weed control is necessary to reduce the competition for soil moisture, nutrients and light between weeds and newly-planted seedlings. Weeds are all plants (including crop and pastures) that will seriously compete with young establishing seedlings in their first few years. Competing weeds will suppress or kill young seedlings, particularly during the summer following planting.

Effective weed control is the most fundamentally important aspect of site preparation for tree planting. It is an essential prerequisite for the successful establishment of windbreak trees. Weed control should be total and achieved before planting. If weed control is achieved well in advance, rain which falls is able to infiltrate and a reservoir of soil moisture develops in the soil profile. This provides greater moisture for early growth and allows planting to be done at times which would otherwise be too dry.

The two primary approaches to weed control are cultivation and the application of herbicides.

Cultivation

Cultivation can provide adequate initial control, but weed regrowth can be rapid and severe. Cultivation can success-fully be used for initial weed control to achieve moisture conservation. Weed seeds which rapidly germinate should then be controlled by herbicides prior to tree planting. Herbicide application involves no soil disturbance, resulting in generally less weed regrowth.

Mouldboard ploughing can produce very high quality weed control through the total inversion of the sod, burying weed seeds. Mouldboarding can be particularly effective on wet sites where mounds can be created by ploughing beside the previous pass in the reverse direction. Planting on mounds in wet, waterlogged areas is generally very beneficial to tree establishment and growth.

Herbicide Spraying

The preferred method for weed control is usually herbicide spraying. *It is important to always read the label provided with any herbicide product and to carefully follow these instructions.*

The critical area to achieve a high level of weed control is the zone within 0.6–1 m around the base of the seedling. Usually strips or spots 1.2–2 m wide are sprayed or alternatively the entire area can be sprayed out. A variety of spray equipment is suitable, including hand held knapsack spray units, spot guns and controlled droplet applicators or boom sprays operated from four-wheel motorbikes or tractors. Spraying strips and spots have some advantages over complete elimination of weeds from the planting area. The amount of herbicide used is reduced. The surrounding weeds and grasses provide some protection to the developing seedlings. The chances of significant soil erosion are less. Also, rabbits and hares are less likely to find and browse seedlings established in sprayed spots.

Either knockdown or a combination of knockdown and residual herbicides can be used for pre-planting weed control. When knockdown herbicides (such as glyphosate) are being used alone, two applications several months apart (say in late autumn and early spring for spring planting) are more effective that a single pre-planting application. Knockdown herbicides should be applied when the weeds are actively growing. The inclusion of a residual herbicide (such as simazine) can prolong the effectiveness of weed control.

Figure 9.1 *Weed control greatly aids seedling growth and survival.*

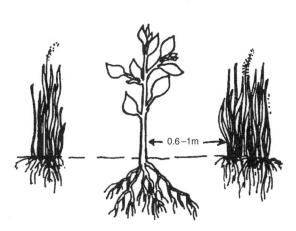

0.6–1m

Residual herbicides usually affect germinating seeds or enter plants through the roots. Care is required with selection and application of residual herbicides in sandy or waterlogging prone soils where they can leach and move from the point of application. The use of a combination of knockdown and residual herbicides can achieve weed control for one year after planting. A variety of additives such as adjuvants, penetrants and wetting agents are also available to improve the effectiveness of herbicides.

When using sprays, apply them on calm days to guard against chemical drift to non-target areas. Rain within 12 hours of spraying can reduce the effectiveness of the herbicide and follow-up treatments may be needed.

It is important to maintain weed control from planting through the first summer of the seedlings' life, when competition for soil moisture is likely to be most extreme. Good pre-planting weed control minimises the need for post-planting weed control.

If post-planting weed control is necessary, this can be achieved by careful application of knockdown herbicides, taking care to shield the seedling from the spray. Long grass sprayed with a knockdown herbicide can brush against trees transferring herbicide to them. Avoid this by cutting the grass before spraying or by flattening it down at the time the herbicide is sprayed. Alternatively, knockdown herbicides can be applied with a wick wiper, although this is time-consuming and tedious for many kilometres of treelines.

In some instances, the seedlings can be over-sprayed with a herbicide which selectively kills the weed but not the seedlings. For example, several herbicides are specific to grasses, but not broad-leaved plants. Over-spraying with residual herbicides which do not affect established plants can also extend the effectiveness of already good weed control.

Suppression of weed regrowth with inert mulches such as woodchips, gravel or woven plastic also extends the effectiveness of initial weed control, but the cost of purchase and placement of these is often prohibitive for large scale plantings.

Weed competition leads, at worst, to the death of the seedling, or at best its stunted growth. Maintaining weed control greatly reduces or eliminates the competition for soil moisture and nutrients in the important first few years of the life of a tree.

Ripping

Some areas have a shallow depth of soils or a layer such as a heavy clay subsoil which limits root penetration. Tree establishment and growth in these situations is greatly assisted by deep ripping or subsoil cultivation. Ripping will provide benefits to tree establishment on most sites, except perhaps for deep sands, deep extremely well-structured soils and cracking clays. Ripping is difficult or impossible in rocky areas, and may produce little benefit on these sites anyhow. Ripping is particularly well suited to the establishment of windbreaks. Long riplines are easily established and they subsequently indicate the location of the rows for planting. Ripping is generally an excellent investment in maximising the success of windbreak establishment.

Deep ripping is usually done with one or more ripping tines drawn by a large tractor or bulldozer. Deep ripping creates a disturbance in the soil profile. The depth of the ripping depends on the pulling power of the machine and the size of the tine. Ripping can be done to any depth up to about one metre and should be at least 30 cm deep to provide any worthwhile benefits. Ripping should be done in dry conditions to achieve maximum shattering of the soil profile. Ripping at this time when the ground is hard and dry is the most difficult, but it is also the most effective. Soil shattering encourages the development of the root systems in three dimensions. Ripping with a single tine can create some problems, particularly when the soil is too wet. No soil shattering occurs with wet soils where the tine simply slices through the soil profile. Two-dimensional root development occurs along the ripline resulting in poor tree growth and stability.

Shattering is increased by the use of winged ripping tines or double ripping with either two tines approximately 50–60 cm apart or using two passes of a single tine at this spacing. The site can also be cross-ripped with the riplines intersecting at right-angles. Cross-ripping is generally better suited to woodlots and plantations than windbreaks.

Riplines will often initially be rough and cloddy and sometimes have air pockets under the surface. It is a good idea to leave the riplines for a few months to allow the lines to settle or to drive along the lines with a tractor wheel to press them down.

Seedlings are planted between double riplines or at the ripline intersections with cross-ripping. Where a single ripline is used, trees should be planted on the shoulder of the ripline rather than right in it. On wet sites, trees planted directly in riplines can become waterlogged in winter.

Ripping of planting lines provides a number of important benefits. Seedlings have a much greater soil volume to exploit due to the shattering of the soil profile. They establish much more rapidly and early growth is increased. Ripping well before planting also allows better moisture infiltration. A reservoir of soil moisture is conserved in the riplines, enhancing growth in dry areas.

Trees established in ripped lines are more wind-firm and are deeper rooting, competing less with neighbouring crops and pastures. Planting of trees is also obviously easier in the loosened, disturbed soil near a ripline.

Vermin Control

Young seedlings can be browsed, particularly by rabbits, hares and native herbivores. It is important to assess the hazard presented by pests prior to establishing seedlings. Some tree and shrub species are far more attractive to herbivores than others. For example, she-oaks and Callitris pines are very palatable. Protection of these from browsing animals is generally essential where they occur.

It is best to control the populations of rabbits with an integrated control program incorporating biological control, poisoning, harbour destruction, and fumigation. To be successful, this approach usually requires a coordinated effort from neighbouring landholders. Attention to detail and vigilance is required as even a small number of rabbits can cause damage. If control of rabbits is not possible, other approaches include erecting vermin-proof fences, individually guarding seedlings with tree guards, or possibly the use of repellents. For long, narrow windbreaks, the use of tree guards is usually more cost effective than vermin-proof fencing. Protection of seedlings for 8–18 months (depending on the rate of growth) is usually sufficient to take them out of reach of rabbit and hares.

Where protection from rabbits is required, tree guards are often the best approach. A wide variety of rabbit-proof tree guards is available. The most commonly used are plastic mesh guards used with two stakes, and plastic sleeves requiring three stakes. Plastic sleeve guards provide the additional benefits of a degree of frost protection and a faster early growth due to a mini "greenhouse" effect.

Plants

Make sure that the planting stock used when establishing windbreaks is the highest quality possible. This means plants propagated and grown to the highest standards and from the best quality genetic material. Try to ensure that the seed used to grow windbreak seedlings was collected from parent trees of superior form. Planting clonal stock from parent trees of known outstanding shelter characteristics is another way to ensure high genetic quality. Although in its infancy, tree breeding to produce seedlings with good shelter characteristics is likely to increase.

A good nursery should be selected and plants ordered well in advance so that you can be sure you will get exactly the seedlings you require. You can collect your own seed and have it propagated for you. This is particularly desirable if you want the indigenous trees to your area or seedlings from parent trees which display particularly desirable growth characteristics. In general, plants should be ordered at least six months ahead of the time you wish to plant. As a rule of thumb, this often coincides with the time you should commence weed control. It is also possible to propagate your own seedlings.

Generally, three kinds of seedlings are available for planting windbreaks: bare rooted, container stock and cell stock.

Bare rooted seedlings are grown in nursery beds where they are root pruned to encourage a fibrous root system. Just prior to sale they are removed soil-less from the bed, bundled and placed in plastic bags. Bare rooted seedlings must be kept cool and moist, and handled carefully. They should be planted quickly after they are received. Traditionally *Pinus radiata* seedlings have been grown bare rooted. In some areas, bare rooted native species are also available. Bare rooted seedlings have the advantage of avoiding the root deformities which can occur in container grown stock and they are generally reasonably priced.

Plants grown in containers are a tried and proven approach. Small plants in 150 mm tubes are preferable to more advanced plants in larger containers. Small plants establish more quickly and have higher survival rates than larger ones. They are also cheaper. Container grown plants should have a deep root system relative to foliage height. In many respects, the quality of the root system is more important than the foliage. Don't be put off by small sized seedlings in tubes. Instead, beware of large luxuriant seedlings with long stems in small containers. Very often their roots will be bound, twisted and constricted. Dig away 2–3 cm of soil from the top of a couple of containers to see if the tap root is distorted or curled. Plants whose roots are bound or distorted should be discarded as they will have poor growth. It is simply a false economy to plant them. Tubes with fluted sides and grown air pruned in raised frames tend to have fewer root problems. Container stock tends to be more expensive, although nurseries can hold these for a longer period of time and therefore plants are more likely to be available in containers if they haven't been ordered.

An increasing number of nurseries are providing cell stock. Cell stock are raised in rectangular trays of usually more than 100 cells which contain the potting material in which the seedling is grown. A range of cell sizes are available. These trays are raised and each cell perforated so that the roots are air pruned. The seedling is removed directly from the tray prior to planting. They are planted quite small and their root development is usually good. They are reasonably priced and establishment can be excellent given good site preparation. Generally cell stock have to be ordered in advance.

Seedlings should be "hardened off" in the few weeks before planting, particularly if their foliage appears soft and luxuriant. This involves reducing water and fertiliser and placing them in a more exposed location. Seedlings should also be well watered on the day they are planted.

Fertilising

Fertiliser is often not necessary. A good idea is to trial fertiliser on just a few seedlings to see if there is a noticeable effect. Fertiliser is usually only required on poor, infertile sites. If fertiliser proves beneficial, complete slow release fertilisers are generally recommended. Make sure that weed control is excellent prior to fertilising, otherwise, you are fertilising weeds as well as seedlings. Fertiliser should never be placed against the tree roots. Pre-planting fertiliser can be placed in the bottom of the planting hole and covered with a little soil prior to planting. For post-planting fertilisation, placing the fertiliser in a spade slit beside the tree is often best. The application of gypsum to the soil prior to planting can benefit seedling establishment on heavy cracking clay soils.

Planting Time

Ideal planting time varies with the region and the season. In wetter areas in southern Australia, more flexibility in planting time is possible. Here, planting later in spring is popular as it enables high quality weed control, and there are fewer frosts and waterlogging problems. In drier areas, planting in winter is usually recommended. Bare rooted stock are usually planted when they are dormant in winter.

Planting

The aim of planting is to place seedlings so that there is good soil contact and that there is minimal distortion and damage to the root system. Generally plant the seedlings about one centimetre deeper than the soil level in the container.

When planting by hand, a number of tools can be used. On hard, difficult ground or sticky clay, a mattock is often the best implement. Bare rooted stock are best planted with a special planting spade which opens up the planting slit allowing insertion of the seedling after which it is firmed in.

For container stock on better loamy soils, specialised treeplanters, such as the Hamilton® treeplanter perform well. This removes a core of soil exactly the same size as the container. Container stock should be well watered prior to planting. The container should be gently removed, minimising root disturbance, and the root ball slipped into the planting hole, before gently firming in. This style of treeplanter is particularly suited to areas where residual herbicides are used, as it does not mix the soil, minimising the chance of the herbicide coming into contact with the tree roots. Two people can plant up to 800 seedlings a day with a Hamilton treeplanter.

Most cell stock systems have a special planting tool designed for them, such as the Pottiputki®. The seedling is placed in a hole the same size as the rootball, and is dropped in through a tube, eliminating the need to bend over. Two people can plant up to 2000 seedlings a day on well-prepared ground using these systems.

The long straight planting lines of many windbreaks on flat terrain is also ideally suited to the use of tractor-mounted planting machines. Mechanical planters generally open a slot in the soil with a tine into which the seedling is dropped. This is followed by press wheels which firm in the seedling.

Many machines can also apply water and fertiliser at planting. Some machines require an operator to place the seedling, while others have a chain drive which carry the seedling into the slot at constant intervals. Different planting machines are available which can handle all of the above kinds of planting stock. Mechanical planting of seedlings is particularly effective for large scale plantings. Windbreaks with long straight runs are ideal. Two people are usually required to operate mechanical tree planters and they can plant anywhere from 2000 to 15 000 seedlings per day. Mechanical tree planters perform best on sites which have been very well prepared.

Watering in may be beneficial, particularly when planting into drier conditions. A litre of water or so should be slowly applied around the seedling. In most areas, no further watering should be needed, given good site preparation. In arid areas, or in very dry summers, watering through the first summer is usually necessary.

Staking of seedlings which are planted when small is generally unnecessary, even in windy situations.

Direct Seeding

Direct seeding is the establishment of trees by sowing seed directly on to an area. It offers a viable alternative to the planting of seedlings for windbreak establishment on many sites.

Direct seeding has several advantages over planting of seedlings for the establishment of windbreaks. In most instances where direct seeding can be reliably undertaken, the cost of windbreak establishment is lower. This is primarily due to the lower labour requirements and because seed is usually cheaper than seedlings. There are also other benefits. Direct sown trees and shrubs usually have better developed root systems than planted seedlings as they are not transplanted during their development. Direct sown windbreaks look much more natural and informal. A range of species can easily be incorporated in the seed mix.

The principle drawback of direct seeding is that failure is more common than with planting. This is because of the great vulnerability of the plants in their early stages of germination and emergence to climatic extremes, fungal, animal and insect pests, and weed competition.

Initial seedling density is very difficult to control with direct seeding. This can be a significant problem for the establishment of narrow windbreaks where uniformity of spacing is important. Consequently, thinning and restocking are more often required. Direct seeding is generally most appropriate for the establishment of wide windbreaks. The broad sugar gum windbreaks in western Victoria, which are still very effective today, were direct sown by far-sighted farmers in the early part of this century.

Direct seeding is best used for species which have rapid emergence and growth from seed. Tree and shrub species which are difficult to germinate and establish are usually best planted as seedlings. A common approach in the establishment of windbreaks is to direct seed one or two rows of fast growing species which are very reliable to establish when sown (such as acacias). Behind these, species which are less reliable to sow and slower to establish are conventionally planted as seedlings.

Direct seeding also relies on a good availability of reasonably priced or easily obtained seed. Where seed of the desired species is particularly valuable or scarce, again, planting of seedlings should be preferred. Note that seed of some species such as most acacias and legumes requires treatment prior to sowing.

There are many different approaches to direct seeding, but all follow the same basic principles. This is not a comprehensive guide to direct seeding as a tree establishment technique. Readers interested in obtaining more detailed information on direct seeding should refer to the book *Direct Seeding of Trees and Shrubs* by Greg Dalton, or the relevant state government extension and advisory service.

The principles of site preparation for direct seeding are similar to planting. High quality site preparation is critical for success. In particular, failure of direct seeding can often be explained by poor weed control. It is extremely desirable to use weed control techniques which provide an extended period of control. Sites may also be ripped in a similar fashion as for planting. Take care when ripping for direct seeding not to bring to the surface infertile subsoil material which can inhibit germination.

The most successful site preparation techniques for direct seeding of windbreaks are the use of herbicides and mouldboard ploughing. Achieving weed control well before sowing time is important to allow the accumulation of a reservoir of soil moisture. This greatly increases the success of direct seeding in dry conditions.

Mouldboard ploughing inverts the ploughed sod, providing long-term control by deeply burying weed seeds.

Knockdown or contact herbicides are usually applied in a number of sprays in strips 1.2–2 m wide over a period of three to six months prior to sowing. Residual herbicides can also be used for direct seeding. They are usually applied in the final spray before sowing. Residuals prevent seeds from germinating by acting as a "chemical blanket". Where residuals are to be applied, it is important to use sowing techniques which ensure that the herbicide sprayed soil does not come into contact with the germinating tree and shrub seeds, as they will suppress their germination as well.

For reliable seed germination on most sites, some form of cultivation or soil disturbance is important. Disturbance is only absolutely necessary in a strip 10–20 cm wide along sowing lines. Many specialised tree seeding machines create this disturbance and sow the seed in the one pass.

Seed may be sown onto the prepared site either by hand or machine. Hand sowing is quick, easy and cheap. The tree and shrub seed is usually bulked out with sand, chick crumbles, bran, or similar inert material before hand broadcasting. It works well for cultivated weed control areas such as those achieved with mouldboard ploughing. Hand sowing can also be used in areas inaccessible to machinery where niche sites have been prepared.

For machine direct seeding, specialised tree seeders work very well and simplify the sowing process. A wide range of commercially produced tree seeders is available. Most scalp or cultivate a narrow sowing line into which tree and shrub seed is precisely placed on the surface at an easily controllable sowing rate. The sowing of long, parallel tree lines by machine is ideally suited to the establishment of windbreaks. Tree seeding machines are best suited to sites prepared by a thorough herbicide spraying regime.

The seeding rate needs to be calculated from the number of viable seeds per kilogram of the species that are being sown and the expected strike rate.

The ideal sowing time is largely dictated by rainfall. In southern Australia, earliest sowing time is generally May for the driest areas (<300 mm rainfall) while the wettest areas

(>900 mm) can be sown as late as October in a normal season. It is generally better to err on the side of sowing a little too early than a little too late.

Windbreak Fencing

This section does not attempt to provide a comprehensive guide to farm fencing which is a major subject in itself. The basic design principles for fences for windbreak protection are emphasised here.

Where livestock are present, even for short periods, windbreaks generally require excellent fencing. In most areas, good windbreaks and good fencing go hand in hand.

Two basic approaches to windbreak fencing exist. Conventional fencing provides a physical barrier to stock, while electric fencing produces a psychological deterrent. Windbreak fences need to be better than normal subdivisional fences since they will be placed under great pressure at times of low feed availability. The exact standard of fencing depends on the kind of stock being excluded. For example, fences excluding hungry cattle or cross-bred sheep need to be stronger than those for passive fine wool merinos. Don't forget that the fence should be designed to withstand the greatest pressure it is likely to be placed under.

Conventional Fencing

The major components of a conventional fence are the end assemblies which anchor the fence, posts which hold the wires in the air and droppers which keep the wires separated between the posts, and of course the wires themselves.

Well-constructed end assemblies are essential for sound fencing — these three design aspects are most important.
- End assemblies are most effective when they incorporate a long horizontal stay at least 3.3 m long, and include a diagonal wire brace.
- Depth of the strainer post is more important than its diameter. A small increase in depth of set from 0.75 m to 0.9 m doubles the load-bearing capacity. Strainer posts should generally be set at least to the depth of their exposed height.
- Driven posts are more stable than those set in post holes.

Three basic wire types are available: plain, barbed, and pre-fabricated. For plain wires, high tensile galvanised wire is long lasting, rust resistant, strong, and maintains good tension. Barbed wire is increasingly rarely used as it is dangerous to sheep and can also be dangerous to install. Prefabricated fences such as Ringlock® and Hingejoint® are effective if well erected, require slightly less work to establish, but are more expensive.

The short strains of some windbreak fences mean that maintaining adequate tension can be a problem. This can be minimised by including mid-line wind-up or spring tensioners.

Electric Fencing

Electric fencing can provide a number of benefits over conventional fencing.

Electric fencing requires wires of much lower tension. The larger number of bends and curves often required in windbreak fencing is more easily accommodated without the need for expensive end assemblies.

A much lighter construction standard is possible with electric fencing as it does not rely on being a physical barrier. This means that electric fences are cheaper and faster to erect. Usually the materials costs are around half that of a comparable conventional fence. They can also be relocated much more readily.

Animals learn to fear the electric wires so that they are rarely touched. Stock new to electric fencing should be educated in special training paddocks or yards with electric fences built to a higher standard to ensure they learn about electricity while being physically constrained. Stock should be well educated before summer when fences may need to be turned off due to fire danger.

The number and arrangement of electric and earth wires depends on a range of factors including the type of stock being excluded, education of stock, management approach and economics. Designs can vary from one hot wire which can be sufficient for dairy cattle, to six closely spaced alternative hot and earth wires required for aggressive, poorly trained sheep. A common design is four alternating earth and hot wires starting with an earth wire near the ground.

Electric fences should be constructed from the highest quality materials to minimise maintenance problems and to ensure long-term effectiveness.

Make sure that the energiser is powerful enough for the length of fence you wish to electrify and that good earthing is provided. Break up the layout into sections with cut-out switches so that fault finding will be easy. It is also desirable

Figure 9.2 *An effectively established windbreak — good fencing, weed control and protection, with seedlings planted into ripped lines. (Photo Andrew Campbell.)*

to be able to isolate different sections of the fencing system when required.

Gates should also be given careful attention to ensure they are not the weak link in the fence. A good standard fence with a slack "cockie's" gate is a false economy. The use of high tensile wires and electric fencing enables the use of cheaper options than traditional gates such "off set lift" gates and "lay down" gates.

Artificial Windbreaks

Artificial windbreaks mostly consist of woven fabric or occasionally timber or metal sheet material attached to a frame or fence. A wide variety of approaches is taken to their construction.

Artificial shelter provides some advantages over conventional natural shelter.

- It can be erected quickly and provides instant shelter, whereas natural shelter can take years to become effective.
- It takes up less space and does not compete with crops for soil moisture.
- Designing the windbreak is more precise, as the height and porosity can be exactly achieved.
- Artificial shelter requires lower maintenance than natural shelter in horticultural applications, except possibly after severe storms when damage to the windbreaks must be repaired.

Artificial windbreaks also have disadvantages.

• The major disadvantage of artificial shelter is its cost which is significantly higher than for natural windbreaks. Intensive artificial shelter in horticultural areas costs in the region of $20–30 000 per ha. The fact that these huge investments are made demonstrates the benefits that windbreaks provide to these high value crops.

• It is probable that greater turbulence is created by artificial windbreaks due to their rigidity.

• Artificial shelter also obviously does not provide the nature conservation or aesthetic benefits of natural windbreaks.

Very often a combination of natural and artificial windbreaks is most effective in horticultural areas. For example, natural windbreaks may be planted around the perimeter of an orchard area and artificial windbreaks used within the orchard, extending the effectiveness of the artificial shelter.

Design and Construction of Artificial Windbreaks

Design and construction of artificial shelter is a specialised task. It is recommended that experienced design consultants and construction contractors are used if investment in an extensive artificial windbreak system is being considered.

Artificial windbreaks are usually constructed with fabric material attached to a fence. Preferably the fabric covers the full height of the fence. Sometimes a design called "overruns" is used where a gap (about 2–2.5 m high) is left between the bottom of the fabric and the ground. Over-runs are not as effective as full height windbreaks and are not recommended for perimeter shelter because of the wind funnelling they cause, which creates accelerated winds for a short distance in the lee. They are sometimes used within a perimeter of full-length windbreaks. Over-runs have the advantages of allowing better access, reduced disruption to crops, and reduced costs. Overall however, full-length shelter is recommended wherever possible.

A wide variety of fabrics is available for artificial windbreaks. They are available in a range of porosities and may be knitted or woven. Knitted fabrics offer some advantages in terms of reduced fraying and being easier to attach to posts and wires.

Sometimes end assemblies for artificial shelter consist of diagonal guy wires or raker poles which anchor the last vertical post. These have the disadvantage of providing an obstruction beyond the fence, but are cheaper to install. It is preferable for end assemblies of artificial windbreaks to have a horizontal stay and a diagonal brace. These provide no obstruction and have greater load-bearing capacity.

The strainer post height required is determined by the depth of set, and the windbreak height. In addition to depth of set, strainer post diameter also affects load-bearing capacity. Either treated softwood or durable hardwood posts can be used. High tensile steel cables or wires are then strained between end assemblies.

Intermediate posts can be free standing in which case they must be set deeply (1.2–3 m) and have reasonable diameter and strength to withstand strong loads. Alternatively, posts may be set less deeply, but should then be guyed and anchored. Lower diameter posts can be used, making this approach cheaper, but the guy wires and anchor points can be an obstruction.

Sometimes no intermediate poles are used and the fabric is attached to the cables strained between end assemblies.

Erection of artificial windbreaks is a specialised job. The fabric material must be firmly attached to one end assembly and stretched into place. A winch or block and tackle can be useful. Most fabrics can be carefully sewn together with black plastic twine, either end to end or lengthwise. A variety of specialised products is available to attach the fabric to posts, pipe frames, and to wires.

CHAPTER 10

Maintaining Your Windbreak

Keeping Windbreaks Working for You

A little ongoing care and maintenance will be repaid by the provision of shelter for many decades. Unfortunately, many well-designed and established windbreaks fail through lack of maintenance. Windbreaks are all too frequently planted and promptly forgotten about. To ensure windbreaks achieve and maintain optimum performance and function, they need to be actively managed.

Replacement Planting and Renovation

A percentage of trees planted will inevitably not survive, especially in their first year. The presence of gaps can be minimised by replacing these losses as soon as possible. This can be in the following planting season at the latest or even in the following summer accompanied by plenty of watering in. Undue delay will result in uneven height within rows, and difficulty in establishing trees in the gaps due to competition with neighbouring trees.

As plants with a shorter lifespan die out, they may require replacement, particularly if they are an important element of the windbreak. Establishing new plants amid or beneath others can be difficult, and planting new rows may be required.

Inevitably every windbreak reaches the end of its functional life. Its replacement should be planned for to ensure continuity of shelter and to minimise costs and disruption.

It is sometimes possible to extend the useful life of a windbreak as it declines in effectiveness through old age. This can be achieved by pruning or lopping declining old tree crowns to encourage new, denser foliage. Seek advice about this, as not all tree species will respond in this way.

More common is the complete felling of trees to allow regrowth from cut stumps (called coppicing), or to allow replanting. Eucalypts in particular often coppice very well. Species which coppice are very useful in windbreaks as they simplify windbreak replacement and renovation. Make sure trees are cut as close as possible to ground level to ensure the development of healthy strong shoots. Trees should not be coppiced in late autumn or winter in cold climates, as the young shoots can be sensitive to frost damage, nor in the middle of summer. It is essential that coppicing stumps be well fenced from stock to enable the new stems to develop. Stock or rabbit browsing can quickly kill the cut stump.

At each coppicing, a small proportion of cut stumps will not recover. Despite this, it is likely that a windbreak of suitable species can be coppiced successfully at least five times.

Coppicing can be a useful method of windbreak renovation for species which have lost their lower foliage (for example, self-pruning eucalypts). The classic example of this applies to the broad windbreaks of sugar gum (*Eucalyptus cladocalyx*) in south-eastern Australia. A quarter to a half width of these belts can be harvested for timber, with the coppicing stumps producing lush new low shelter. When these are sufficiently established, another strip can be harvested, and so on.

Timberbelts usually have a planned life and replacement strategy, determined by the ideal harvesting time for timber. The timber generates revenue which should be partially reinvested in the re-establishment of the new windbreak. Some shelter can be maintained by the low shelter rows, while the new timber rows are establishing.

Figure 10.1 *Coppicing showing correct and incorrect stump heights.*

The correct and incorrect stump heights

Weed Control

Weed control is the most important windbreak maintenance practice immediately after tree and shrub establishment. Maintaining a weed free area 1.2–2 m in diameter for 12 to 18 months after planting minimises losses and increases tree and shrub growth. This can be achieved by application of knockdown herbicides on emerging weeds, taking care to shield the young trees. Alternatively, knockdown herbicides can be carefully applied to the weeds with a wick wiper. Careful use of appropriate residual herbicides (either applied in combination with knockdown herbicides for pre-planting weed control, or over-spayed after planting) provides extended weed-free periods around shelter trees. Even after the windbreak is well established, tree and shrub growth will be enhanced by continued weed control.

Fences

All windbreaks in grazing areas should be permanently fenced and those fences maintained in good condition. Unfenced windbreaks will have no foliage up to browse level, exactly where it is needed the most. This situation can be a death-trap for stock near the windbreak where wind funnels at high speed through this gap. In addition, some tree species, such as the stringybark group of eucalypts, are susceptible to stock damage and death by stock browsing and rubbing.

Tree fences must be maintained to a very high standard as they will be placed under great pressure when feed is scarce and the treed area has been ungrazed. One of the most common situations where trees are damaged by stock on farms is when a new fence is erected beside an existing fence with a windbreak developed between them. The extra pressure suddenly placed on the old fence causes it to fail, resulting in tree losses. Stock sheltering from wind and sun near windbreaks can also place extra pressure on non-electric fences by rubbing on fence posts and droppers.

Pest Control

Insect pests in establishing windbreaks should be monitored and controlled if the infestation becomes severe. Insects are normally kept in check through predation by birds and mammals and by parasites. Occasionally these natural balances are upset, and insects can build up. Light infestations are rarely significant, but repeated severe defoliation affects the health and vigour of the host seedlings. When this occurs, insect control measures can be undertaken. Tree species planted well outside their normal range are more prone to defoliation than indigenous species. The planting of a variety of smaller trees and shrubs can provide habitat for a range of birds, predatory and parasitising insects and mammals which provide natural insect control.

Regular monitoring for insect problems is recommended to enable early detection of potential problems before they become serious.

For Australian natives, leaf-eating insects and sap-ingesting insects are the most common. Common leaf-eating insects are leaf blister sawflies, cluster sawflies (or spitfires), autumn gum moths, christmas beetles, and gum leaf skeletoniser. These cause defoliation of the seedlings to varying degrees. Sap ingesting insects include psyllids (or lerps), aphids, and gum tree scale. These insect problems can generally be removed by pruning affected foliage if the infestation is small or by spraying with a contact insecticide such as malathion. Where the insect is hidden or shielded by a protective covering such as leaf blister sawfly, psyllids or gum tree scale, a systemic insecticide such as dimethoate should be used. These enter the sap stream and are ingested by the insects. The addition of a wetting agent and white oil assists in the control of scale insects. Carefully directed spot applications of insecticides are often better than blanket application as this can kill a range of non-harmful and even beneficial insects.

Red-legged earth mites and wingless grasshoppers can damage young seedlings in some areas. Specific pesticides are available for both of these which give short-term control and can be applied in a band along the planting lines.

All insecticides should be applied strictly in accordance with the instructions on the label. They should be applied only on calm days to minimise spray drift to non-target areas.

In general, insecticides should be applied only where the infestation becomes serious and no alternative control methods are suitable.

Rabbits should be controlled by conventional techniques of poisoning, warren ripping, harbour reduction, and fumigation. Alternatively, trees may be protected by the use of rabbit-proof tree guards. Hares can also be damaging but this can be reduced by shooting or the use of guards. Spot spraying for weed control, rather than strip spraying tends to reduce the severity of losses caused by rabbits and hares.

Agricultural Management

Livestock need to be managed so that they can take maximum advantage of shelter. It's no good having outstanding shelter if sensitive livestock are not there when they need it. Sensitive stock may need to be confined in highly sheltered areas, such as a woodlots, sheltered laneways or livestock havens during critical periods.

Access points should be provided in multi-row windbreaks to allow easy entry for maintenance and management. These access points need not be conventional galvanised steel swing gates, nor should they be "cockie's gates". Lay-down panels for traditional post and wire fencing and lift-up sections for electric fencing can work well. In very severe conditions stock have been saved by being driven into wide windbreaks, even when trees are young. Damage to trees is minimal and more than offset by stock numbers saved.

A common complaint is a decrease in crop or pasture vigour immediately beside windbreaks. Deep ripping with a ripping tine or root pruning with a large rolling coulter severing tree roots outside the limit of tree crowns can limit competition between trees and crops or pastures. This should be done every few years alternately each side of the windbreak to minimise moisture stress and any lack of tree stability.

Pruning

Pruning of trees in windbreaks can be undertaken for many reasons, including to improve the timber quality of trees in windbreaks, to reduce fire risk, to reduce shading of adjacent pastures or crops, and to increase the stability of trees to wind.

Pruning to improve timber quality is the most common form of windbreak pruning. Trees can be high pruned to achieve a clear knot-free butt log by removing lower branches. This produces a much more desirable log for sale, making marketing of small volumes of farm-grown logs far easier. Pruned logs command significantly higher prices than unpruned logs. It is a poor investment to prune trees with little potential to produce timber. It is preferable to thin these poorly formed trees out, provided this will not impair windbreak function.

The most common approach to pruning for timber improvement is termed "variable lift" pruning. This involves progressively increasing the height pruned, commencing at two to three years of age, continuing approximately annually up to 6 m as the tree grows. Generally, all branches are pruned off below a certain tree trunk diameter (often about 10 cm). This will result in logs with the branch stubs confined to a small knotty core. Make sure that not more than 50% of the tree crown is removed, as this will lead to reduced tree growth. Regular pruning is recommended to prevent the development of large branches.

Higher branches can be selectively pruned to pre-empt long-term branching problems. These may be totally pruned or tip pruned. Tip pruning is where branch growth is retarded by the removal of the growing tip and part of the foliage of the branch. Early removal of heavy, large branches from the upper part of the crown as the trees develop will reduce their sail area, and thereby increase their wind firmness.

Pruning should remove the branch leaving the smallest wound possible. Prune the branch as close to the trunk as possible, but do not damage the trunk or branch collar as this will greatly retard the healing and occlusion of the branch wound.

Figure 10.2 *The correct and incorrect way to prune trees for clearwood production (left) and the effect of elimination of knots created by clearwood pruning (right). (Source: Reid & Stewart, 1996.)*

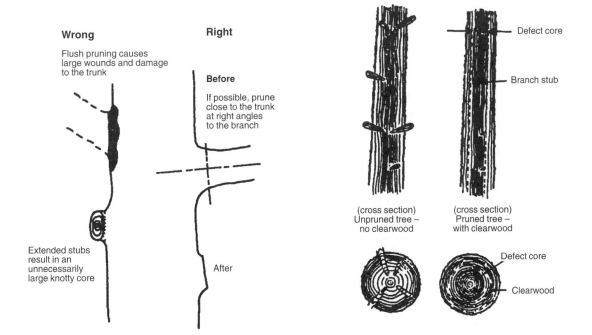

Wrong

Flush pruning causes large wounds and damage to the trunk

Extended stubs result in an unnecessarily large knotty core

Right

Before

If possible, prune close to the trunk at right angles to the branch

After

(cross section)
Unpruned tree –
no clearwood

(cross section)
Pruned tree –
with clearwood

Defect core

Branch stub

Defect core

Clearwood

A variety of equipment is available for pruning, including hand shears, hand saws and power saws, all of which are available with extendable handles. Specially made ladders are also available for high pruning timber trees.

Timber trees in windbreaks can also be pruned to improve their form if this is poor. Form pruning is generally undertaken on young trees, up to two years old. It involves encouraging a strong dominant leader stem by removing competing leaders, large secondary branches, and branches at an acute angle. The desired result is the development over time of a straight single trunk with small branches. Form pruning is essential to maximise timber production potential where the initial planting density is not sufficient to simply select the best form trees by thinning out the inferior ones.

Obviously, when high pruning timber rows, low shelter must be provided by other rows in the windbreak. An alternative approach is fan pruning where branches perpendicular to the windbreak are removed. In this way, clear timber is produced on two sides of the tree while shelter is maintained.

Another innovative pruning variation aiming to provide protection of the trunk of the tree while producing high value clearwood is the "poodle" cut. Here branches below about 0.5 m are retained to prevent stock from accessing the trunk, while branches above this are pruned in the normal manner.

Side Trimming

Some tree species with broad spreading crowns may eventually develop branches extending over the fenceline. This will occur in particular where insufficient space has been left between the outer windbreak row and the fence. To minimise damage to fences by falling branches, trees in the outer rows can be pruned. Where land is particularly valuable, such in horticultural areas, the spacing from the outside windbreak rows to the fence can be reduced and a program of regular pruning undertaken to minimise growth of branches and foliage across the fence and into the paddock.

Thinning

Trees may need to be removed (or thinned) from windbreaks from time to time, especially if a dense initial spacing has been used to provide early shelter. Thinning can be used to increase the porosity of very dense windbreaks.

It is a sound approach to plant timber rows of timberbelts at a close spacing to enable thinning and to encourage the development of straighter, better formed trees. This allows for the removal of trees with poor form. Only the best performing trees for timber production should be allowed to grow on. Planting high quality planting stock selected for its form and timber value reduces the level of thinning required for a successful final crop. The use of high quality radiata pine clonal stock, for example, can virtually eliminate the need to thin. Seedlings of many native species are totally unimproved from a timber perspective and higher selection ratios are necessary to achieve a high quality timber crop. In these cases, to achieve the highest quality timber, the poorest half to two-thirds of the trees in timber rows may need to be removed. This means initially planting double or triple the final intended tree density. These higher initial planting densities also provide faster early shelter.

Bibliography

Aase, J K & Siddoway, F H (1976), "Influence of tall wheat-grass barriers on soil drying", *Agron J* 68: 627–31.

Anderson, G (1987), "The effect of trees on crop and animal production", *Trees and Natural Resources,* vol 28, No 3: 14–17.

Baldwin, C S (1988), "The influence of field windbreaks on vegetable and speciality crops", *Agric Ecosystems Environ* 22/23: 191–203.

Benzarti, J (1990), "Effects of windbreak on microclimate and agricultural production in Tunisian irrigated lands", *Protective Plantation Technology,* ed Xiang Kaifu, Shi Jiachen, Baer, N W & Sturrock, J W , Northeast Forestry University, Harbin, pp 218–24.

Bird, P R (1991), "The role of trees in protecting soils, plants and animals in Victoria", *Papers of The Role of Trees in Sustainable Agriculture, a national conference*, Rural Industry Research and Development Corporation, Canberra, pp 77–87.

Bird, P R, Lynch, J J & Obst, J M (1984), "Effect of shelter on plant and animal production", *Proc Aust Soc Anim Prod* 15: 270–3.

Black, A L & Aase, J K (1988), "The use of perennial herbaceous barriers for water conservation and the protection of soils and crops", *Agric Ecosystems Environ* 22/23: 135–48.

Boylen, D E (1983), "Land Degradation in the arid zone of Queensland", *Proceedings of the seminar, Agriculture and Conservation in Central Queensland*, Dalby Agricultural College.

Brandle, J R , Johnson, B B & Dearmont, D D (1984), "Windbreak economics: the case of winter wheat production in eastern Nebraska", *J Soil Water Conserv* 39: 339–43.

Brown, K W & Rosenberg, N J (1972), "Shelter-effects on microclimate, growth and water use by sugar beets in the Great Plains", *Agric Meteorol* 9: 241–63.

Burke, S J A (1991), "Effect of shelterbelts on crop yields at Rutherglen", *The Role of Trees in Sustainable Agriculture, a national conference*, Rural Industry Research and Development Corporation, Canberra, pp 89–99.

Burke, S J A, Campbell, C A & Robertson, D I (1988), *Shelterbelts*, Department of Conservation, Forests and Lands, Melbourne.

Burke, S J A & Wilson, A (1997), "A haven from storm and drought", Rural Industries Research and Development Corporation.

Caborn, J M (1957), *Shelterbelts and Microclimate*, Comm Bull no 29, Edinburgh.

Caborn, J M (1965), *Shelterbelts and Windbreaks*, Faber & Faber, London.

Carr, M K (1972), "The internal water status of the tea plant (*Camellia sinensis*): some results illustrating the use of the pressure chamber technique", *Agric Meteorol* 9: 42–6.

Cooke, J W, Maclennan, H S & Erlandsen, S A (1989), "Arable farming systems", *Mediterranean Landscapes in Australia*, ed Noble, J C & Bradstock, R D, CSIRO, Melbourne, pp 318–28.

Country Fire Authority (1992), "Tree Planting for Fire Protection on Farms", *Information Bulletin no 009/92*, Melbourne.

Dalton, G (1993), "Direct seeding of trees and shrubs", Primary Industries South Australia, Adelaide.

Davison, T M, Silver, B A, Lisle, A T & Orr, W N (1988), "The influence of shade on milk production of Holstein–Friesian cows in a tropical upland environment", *Aust J of Experimental Agriculture* 28:149–54.

Ding Guifang & Zhang Yuhua (1990), "Effects of shelterbelt network on soil moisture", *Protective Plantation Technology*, ed Xiang Kaifu, Shi Jiachen, Baer, N W & Sturrock, J W, Northeast Forestry University, Harbin, pp 180–202.

Donnelly, J B (1984), "The productivity of breeding ewes grazing on lucerne or grass and clover pastures on the tablelands of southern Australia; III: Lamb mortality and weaning percentage", *Aust J Agric Res* 35: 709–21.

Drake, B G, Raschke, K & Salisbury, F B (1970), "Temperatures and transpiration resistances of Xantium leaves as affected by air temperature, humidity and windspeed", *Plant Physiol* 46: 324–30.

Fitzgerald, D (1994), *Money Trees on your Farm*, Inkata, Melbourne.

Frota, P C E, Ramos, A D & Carrari, E (1989), "Wind profile and soil water availability near a Caatinga shelterbelt", *Meteorology and Agroforestry*, ed Reifsnyder, W S & Darnhofer, T O , ICRAF, Nairobi, pp 531–6.

Garrett, B (1989), *Whole Farm Planning, Principles and Options*, Department of Conservation and Natural Resources, Melbourne.

Grace, J (1988a), "Plant response to wind", *Agric Ecosystems Environ* 22/23: 71–88.

Greb, B W & Black, A L (1961), "Effects of windbreak plantings on adjacent crops", *J Soil and Water Cons* 16: 223–7.

Guyot, G (1989), "Les effets aerodynamiques et microclimatiques des brise-vent et des amenagements regionaux" *Meteorology and Agroforestry*, ed Reifsnyder, W S & Darnhofer, T O , ICRAF, Nairobi, pp 485–520.

Hagen, L J & Skidmore, E L (1974), "Reducing turbulent transfer to increase water use efficiency", *Agr Meteorol* 14: 153–68.

Heisler, G M, & DeWalle, D R (1988), "Effects of windbreak structure on wind flow", *Agric Ecosystems Environ* 22/23: 41–69.

Holmgren, D (1987), *Trees on the Treeless Plains*, Project Branchout, Hepburn.

Huxley, P A, Darnhofer, T, Pinney, P A, Akunda, E, & Gatama, D (1989), "The tree crop interface: a project designed to generate experimental methodology", *Agroforestry Abstracts* vol 2 no 4, pp 127–45.

Judd, M J (1988), "Windbreaks", *Optimising Horticultural Development in the Cool Temperate Environment*, Department of Agriculture, Tasmania, pp 105–25.

Kang, B T (1993), *Alley cropping – past achievements and future directions*, Agroforestry Systems 23: 141–55.

Lefroy, T & Scott, P (1993), "Alley farming", *WA J of Agriculture* 4:119–32.

Lynch, J J & Donnelly, J B (1980), "Changes in pasture and animal production resulting from the use of windbreaks", *Aust J of Agricultural Research* 31: 967–79.

Lynch, J J, Elwin, R L & Mottershead, B E (1980), "The influence of artificial windbreaks on loss of soil water from a continuously grazed pasture during a dry period", *Aust J of Experimental Agriculture and Animal Husbandry* 20: 170–4.

Marshall, J K (1967), "The effect of shelter on the productivity of grasslands and field crops", *Field Crop Abstr* 20: 1–14.

McAneny, K J, Judd, M J & Trought, M C T (1984), "Wind damage to kiwifruit (*Actinidia chiniensis*) in relation to windbreak performance", *NZ J Agric Res* 27: 255–63.

McMartin, W, Frank, A B, & Heintz, R H (1974), "Economics of shelterbelt influence on wheat yields in North Dakota", *J Soil Water Conser* 29: 87–91.

McNaughton, K G (1988), "Effects of windbreaks on turbulent transport and microclimate", *Agric Ecosystems Environ* 22/23: 71–88.

Monteith, J L (1964), "Evaporation and the environment: the state and movement of water in living organisms", *19th Symp Soc Exp Biol*, Academic Press, New York, pp 205–34.

Naegeli, W (1964), *On the most favourable windbreak spacing*, Scottish Forestry, 18: 4–15.

Ong, C K (1993), "The dark side of intercropping: the manipulation of soil resources", *Proceedings of the Ecophysiology of Intercropping Conference, Guadeloupe, 6–12 December 1993*, Paris Institut National de la Recherche Agronomique.

Peck, A J & Hurle, D H (1973), "Chloride balance of some farmed and forested catchments in south-western Western Australia", *Water Resources Res* 9: 648–57.

Penman, H L (1948), "Natural evaporation for open water, bare soil and grass", *Proceedings of the Royal Society London* A193, pp 120–46.

Platt, S (1995), *Including Wildlife in Landcare Actions*, Land for Wildlife note no 30, Department of Conservation and Natural Resources (Vic), Melbourne.

Radcliffe, J E (1985), "Shelterbelt increases dryland pasture growth in Canterbury", *Proceedings of the New Zealand Grassland Association Conference.*

Radke, J K & Hagstrom, R T (1973), "Plant-water measurements on soybeans sheltered by temporary corn windbreaks", *Crop Sci* 13: 543–8.

Radke, J K (1976), "The use of annual wind barriers for protecting row crops", *Proceedings of Symposium, Shelterbelts on the Great Plains*, 20–22 April 1976, Denver Co, ed Tinus, R W, Great Plains Agricultural Council Publication no 78, pp 79–87.

Raine, J K & Stevenson, D C (1977), "Wind protection by model fences in a simulated atmospheric boundary layer", *J Indust Aerodyn* 2: 159–80.

Reid, R & Stewart, A (1994), *Agroforestry, Productive Trees for Shelter and Land Protection in the Otways*, Otway Agroforestry Network, Birregurra.

Reifsnyder, W E (1989), "Control of solar radiation in agroforestry practice", *Meteorology and Agroforestry*, ed Reifsnyder, W S & Darnhofer, T O, ICRAF, Nairobi, pp 141–57.

Rosenberg, N J (1966), "Microclimate, air mixing and physiological regulation as influenced by wind shelter in an irrigated bean field", *Agric Meteorol* 3: 197–224.

Rosenberg, N J, Blad, B L, & Verma, S B (1983), *Microclimate, the Biological Environment*, John Wiley & Sons, New York.

Russell, G & Grace, J (1978), "The effect of windspeed on the growth of grasses", *J Appl Ecol* 16: 507–16.

Russell, G & Grace, J (1979), "The effect of shelter on the yield of grasses in southern Scotland", *J Appl Ecol* 16: 319–30.

Schwartz, R C, Fryear, D W, Harris, B L, Bilbro, J D & Juo, A S R (1995), "Mean flow and shear stress distributions as influenced by vegetative windbreak structure", *Agriculture and Forest Meteorology* 75: 1–22.

Siddique, K H M, Tennant, D, Perry, M W & Belford, R K (1990), "Water use and water use efficiency of old and modern wheat cultivars in a Mediterranean-type climate", *Aust J Agric Res* 41: 431–47.

Singh, D & Kohli, R K (1992), Agroforestry Systems, pp 253–66.

Skidmore, E L, & Hagen, L J (1970), "Evaporation in sheltered areas influenced by windbreak porosity", *Agr Meteorol* 7: 363–74.

Skidmore, E L (1976), "Barrier-induced microclimate and its influence on growth and yield of winter wheat", *Proceedings of Symposium, Shelterbelts on the Great Plains*, 20–22 April 1976, Denver Co, ed Tinus, R W, Great Plains Agricultural Council Publication no 78, pp 57–63.

Snook, L C (1986), *Tagasaste*, Night Owl (pub), Shepparton.

Sun, D & Dickson, G R (1994), "A case study of shelterbelt effect on potato (*Solanum tuberosum*) yield on the Atherton Tablelands in tropical north Australia", *Agroforestry Systems* 25: 141–51.

Stern, W R (1967), "Seasonal evapotranspiration in irrigated cotton in a low altitude environment", *Aust J Agric Res* 18: 259–69.

Stoeckeler, J H (1962), "Shelterbelt influence on great plains field environment and crops", USDA Prod Res Rep no 62.

Sturrock, J W (1969), "Aerodynamic studies of shelterbelts in New Zealand – 1", *NZ J Sci* 12: 754–76.

Sturrock, J W (1972), "Aerodynamic studies of shelterbelts in New Zealand – 2", *NZ J Sci* 15: 113–40.

Sturrock, J W (1981), "Shelter boosts crop yields by 35 per cent: also prevents lodging", *NZ J Agric* 143: 18–19.

Van Einerm, J, Karschon, R Razuinova, L A, & Robertson, G W (1964), *Windbreaks and Shelterbelts*, WMO tech note no 59, WMO, no 147, TP 70, p 160.

Index